GEOLOGY UNDERFOOT
IN NORTHERN ARIZONA

GEOLOGY UNDERFOOT
IN NORTHERN ARIZONA

Lon Abbott and Terri Cook

Mountain Press Publishing Company
Missoula, Montana
2007

© 2007 Lon Abbott and Terri Cook
All rights reserved

All photographs by authors unless otherwise credited.

Cover art: West Mitten Butte in Monument Valley.
Original painting by Dona Abbott.

The Geology Underfoot series presents geology with a hands-on, get-out-of-your-car approach. A formal background in geology is not required for enjoyment.

**is a registered trademark of
Mountain Press Publishing Company.**

Library of Congress Cataloging-in-Publication Data

Abbott, Lon, 1963–
  Geology underfoot in northern Arizona / Lon Abbott and Terri Cook. —1st ed.
    p. cm.
  Includes index.
  ISBN 978-0-87842-528-0 (pbk. : alk. paper)
  1. Geology—Arizona—Guidebooks. 2. Geoparks—Arizona—Guidebooks. 3. National parks and reserves—Arizona—Guidebooks. 4. Parks—Arizona—Guidebooks. I. Cook, Terri, 1969– II. Title.
QE86.N67A22 2007
557.791—dc22
                              2007004869

Printed in the United States of America

Mountain Press Publishing Company
P.O. Box 2399
Missoula, Montana 59806
(406) 728-1900
www.mountain-press.com

For Dona,
With everything you do
and every life you touch,
you paint a better world.

Dona Abbott at work in her studio

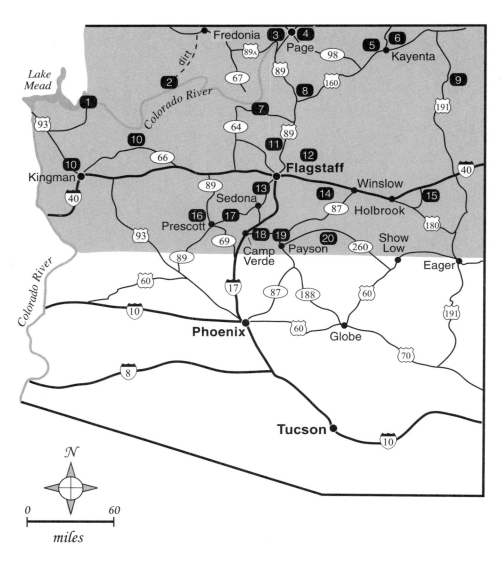

This book covers the part of Arizona that is shaded on the map. Numbers correspond to vignette numbers.

Contents

Preface *ix*

The Geology of Northern Arizona *1*

1. A Grand Transition: Pearce Ferry *9*

2. A Conflict of Fire and Water: Toroweap Overlook *25*

3. Flood of Controversy: *39*
The Colorado River and Glen Canyon Dam

4. Jewel of the Southwest: Antelope Canyon *53*

5. A Tale of Twin Plateaus: *65*
Navajo National Monument and Black Mesa

6. How to Carve a Totem Pole: Monument Valley *79*

7. Journey to an Ancient Shoreline: The Grand Canyon *93*
and the Gorge of the Little Colorado River

8. *Dilophosaurus* Dancehall: Moenave Dinosaur Tracks *107*

9. Beauty All around Me, with It I Wander: *121*
Canyon de Chelly National Monument

10. Volcanic Violence in a Landscape Turned Upside Down: *135*
The Peach Springs Tuff

11. A Mélange of Magmas, a Variety of Volcanoes: *149*
The San Francisco Volcanic Field

12. Desert Niagara: Grand Falls *165*

13. Pangaean Riviera: The Sedona Red Rocks *175*

14. A Crater with Deep Impact: Meteor Crater *189*

15. Triassic Time Capsule: Petrified Forest National Park *203*

16. Continent Under Construction: The Rocks of Prescott *217*

17. Mountain of Metal: The Mines of Jerome *233*

18. A Lake's Legacy: Montezuma Castle National Monument   *249*

19. A Bridge to the Past: Tonto Bridge State Park   *261*

20. Islands in Time: The Mazatzal Quartzite   *271*

Glossary   *285*

Sources of More Information   *297*

Index   *305*

Preface

The geologic diversity on display in northern Arizona is unsurpassed anywhere on earth. Introductory geology textbooks are filled with photographs of destinations in this area, such as the Grand Canyon, Monument Valley, the Petrified Forest, and Meteor Crater—all world-class examples of geologic features and processes. Less celebrated attractions in the region are no less fascinating. This book is a guide to twenty noteworthy sites in northern Arizona and the extraordinary events that have sculpted this spectacular landscape over the last 1,800 million years.

We have written this book so that each of the twenty vignettes is a self-contained story, yet together the vignettes constitute a biography of northern Arizona through the vastness of geologic time. For the reader who wishes to trace northern Arizona's remarkable evolution, we have compiled a table with which you may follow vignettes chronologically, transforming your travels through the region into a journey through time. Additional sources of information listed at the back of the book enable you to delve deeper into any of the topics.

The story of the earth is told in the language of rocks. A crucial piece of this vocabulary is the theory of plate tectonics, and if you're not familiar with this concept, we urge you to read the first few paragraphs of vignette 16, which discuss the theory in the context of the early formation of northern Arizona's crust.

Unless we say otherwise, dirt roads we send you on may be bumpy or dusty, but they should be passable in good weather in a passenger car with normal clearance. In wet weather, some roads become impassable. Check road conditions locally, prepare, and don't go if in doubt. In remote areas and along some roads, no services are available, so make sure you have a good spare tire and plenty of water, gasoline, and food. Exercise extreme caution clambering around on rocks and near cliff edges. And never enter a slot canyon if there's a chance heavy rain could fall on any part of the canyon's drainage area. It need not be raining on you for a flash flood to reach you.

We are indebted to a number of people who helped this project come to fruition. Dona Abbott's beautiful artwork on the cover and in illustrations greatly enhanced the book. Bill Ervin offered invaluable help as we learned to use our graphics software. We thank Fred Hanselmann of

TIMELINE OF MAJOR GEOLOGIC EVENTS IN ARIZONA'S HISTORY AND VIGNETTES IN THIS BOOK THAT EXPLORE THEM

YEARS AGO (Ma = million years ago)	MAIN GEOLOGIC EVENTS	VIGNETTE
1,800–1,700 Ma	Formation and amalgamation of volcanic arcs. Collision with proto–North American continent.	16, 17
1,700–1,600 Ma	Erosion of uplifted mountains deposits material for Mazatzal quartzite. Subsequent collision between growing North American continent and another arc deforms the Mazatzal quartzite. (Mazatzal Orogeny)	20
1,400 Ma	Intrusion of granite due to unknown causes.	
1,200–750 Ma	Period of general quiet; sediment deposition in shallow seas. Minor period of extension 1,100 million years ago in response to continent-continent collision that formed the supercontinent Rodinia.	20
750 Ma	Rodinia breaks up, creating series of Basin-and-Range-like mountain chains across northern Arizona.	
750–542 Ma	Erosion of mountain chains reduces northern Arizona to a nearly flat plain except for a few Mazatzal and other resistant islands.	20
542–300 Ma	Quiet interlude with deposition of a thick sedimentary sequence in a series of shallow seas and coastal environments.	17, 19, 20
300–251 Ma	Assembly of Pangaea raises Ancestral Rockies in Colorado. Sediments shed west from these mountains blanket northern Arizona, which lies along the super-continent's west coast. Progressively drier climate results in dune deposition.	6, 7, 9, 13
251 Ma	Mass extinction of 90 percent of Earth's species.	14, 15

YEARS AGO (Ma = million years ago)	MAIN GEOLOGIC EVENTS	VIGNETTE
251–200 Ma	Life recovers; fossils from Petrified Forest display considerable ecological complexity. Pangaea begins to break up, triggering subduction zone volcanoes in southern Arizona.	9, 15
200–146 Ma	Another mass extinction kick-starts dinosaur dominance. Extremely arid climate generates a Saharan-scale desert in northern Arizona.	4, 5, 8
146–65 Ma	Sediment shed from the growing Sevier Highlands to the west blankets the area. Coal is deposited in coastal swamps of an eastern sea.	1, 5, 10
65 Ma	Asteroid impact in Yucatan, Mexico, wipes out 70 percent of earth's species, including dinosaurs.	14
65–30 Ma	Tectonic activity returns to northern Arizona after a long hiatus, raising the Mogollon Highlands to the south and the modern Colorado Plateau.	5, 9
30–5 Ma	Plate reorganization off the California coast stretches the crust, causing the collapse of the Sevier-Mogollon Highlands and creating the modern Basin and Range province. Lakes fill some basins, including the Verde Valley.	1, 10, 18
5 Ma–10,000 years	The Grand Canyon is carved. Expansion of the Basin and Range thins the crust and melts magma. Volcanic fields pop up in several locations across northern Arizona.	1, 2, 11, 12
50,000 years	Asteroid slams into northern Arizona's high plains, blasting a mile-wide crater.	14
10,000 years–present	Springs deposit travertine limestone in river canyons. Human exploitation of geologic and water resources alters landscape and in some places triggers environmental degradation.	3, 5, 17, 19

Meteor Crater Enterprises; Jerome State Historic Park manager Mike Rollins; Wayne Ranney; and Veronique Robigou for permission to use their photographs. E-mail conversations with Martin Lockley and reviews of all or part of the manuscript by Wayne Ranney and Mike Rollins substantially improved the final product. This project would not have been possible without Mountain Press Publishing Company and our supportive and expert editor, Beth Judy, who made writing this book a pleasure.

The Geology of Northern Arizona

Northern Arizona is a land of contrasts. From the plunging depths of the Grand Canyon to jagged volcanic peaks, from Sedona's vibrant red rocks to Antelope Canyon's tapered slot, northern Arizona's magnificent landscapes beckon visitors to experience their stunning beauty and tremendous variety. Etched into its world-famous vistas are stories of sweeping changes the region has endured. The region's rocks record vivid tales of mighty mountain ranges raised and torn down, tropical seas come and gone, vast deserts of sand dunes a thousand feet high marching to the horizon, and plants and animals that have called northern Arizona home. As you travel the roads of northern Arizona with this book as your guide, you can piece together the region's long and tortuous transformation from a nondescript patch of ocean floor to its present splendor. This overview of northern Arizona geology will help you place the geologic features discussed in the vignettes into the broader context of the area's vast 1,800-million-year history.

The earth and the rest of the solar system coalesced from stardust 4,600 million years ago. To deal with such a vast sweep of time, geologists divided and subdivided earth's history into shorter chunks along the timeline from past to present, from eons to eras to periods. There are two eons: the Precambrian, which spans the first 80 percent of earth history, and the Phanerozoic, which is further broken into three eras.

For the first 2,800 million years of earth's existence, the patch of ground that would become northern Arizona was a nondescript scrap of barren seafloor covered by dense oceanic crust destined for obliteration in a subduction zone. No rock record from this early phase remains.

That changed about 1,800 million years ago, when a series of subduction zone volcanoes forged the area's first continental crust. Northern Arizona at that time consisted of a tangle of mountainous island chains somewhat like today's Southeast Asian archipelagos. Vignette 16 narrates the birth and life of one such chain of volcanic islands in the Prescott area. A similar chain comprising the land around modern Jerome (vignette 17) was simultaneously erupting several hundred miles away. Subduction caused the two arcs to viciously collide about 1,700 million years ago.

Northwest of these island chains lay the early North American continent, busily growing by adding material to its edges. The amalgamated

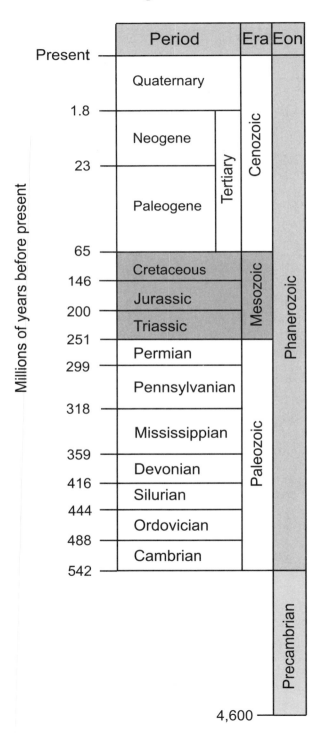

arcs barely had time to recover before North America swept them up like bugs smashing into a windshield. The collision extended the continent's width by 300 miles and raised a chain of mountain ranges that stretched across northern Arizona and as far east as Nebraska. Hot springs associated with the former volcanic islands deposited Jerome's rich copper ores to form a mountain of metal (vignette 17), while unusual iron deposits documenting profound changes in earth's early atmosphere accumulated on the flanks of Prescott's volcanic arc (vignette 16).

Incited by fresh, steep slopes and mountain weather, erosion tore at these highlands, reducing their steep volcanic peaks and tall granite walls to boulders, pebbles, and sand that blanketed the landscape, forming a widespread sediment layer known as the Mazatzal quartzite. Continued erosion of the mountains buried the Mazatzal sediment, which was metamorphosed into a rock of singular durability and soon thereafter contorted when another chain of volcanic islands smashed into the continent 1,600 million years ago—the arrival of southern Arizona (vignette 20).

After all this tectonic excitement and a final injection of granite 1,400 million years ago, northern Arizona settled down. Mountains eroded away, exposing granite and metamorphic roots by 1,200 million years ago. Northern Arizona became a nearly flat plain lying at sea level, with only a few small islands of incredibly resistant Mazatzal quartzite sticking up.

At this time a sea encroached on the region from the south and west, depositing a thick sequence of sedimentary rocks as it waxed and waned across the landscape during the next 450 million years. Near Payson, a series of rock layers called the Apache group was deposited as this sea lapped upon the shores of a Mazatzal quartzite island that has been reexposed and is visible today (vignette 20).

About 1,100 million years ago, this quiet was shattered when basaltic lavas began oozing across the landscape. Northern Arizona was stretching —probably the by-product of a massive continent-to-continent collision that deformed much of North America's current East and Gulf Coasts and uplifted a Himalayan-scale mountain range in present-day Texas. With this culminating event, essentially all of the planet's continental crust was bound together in a supercontinent known as Rodinia. Northern Arizona lay locked in the center.

The earth contains a tremendous amount of internal heat that must be released. Supercontinents trap that heat, which builds up and ultimately forms rift zones that tear them apart. The destiny of all supercontinents is to break up, which is exactly what happened to Rodinia about 750 million years ago. One of the main rift zones lay a short distance west of Arizona near modern Las Vegas, and it caused significant landscape changes across the region. Adjacent blocks of crust, including the layers of the Apache group, were offset almost 2 vertical miles along a series of major faults. Northern Arizona back then looked like southern Arizona does now, with its dominant Basin and Range topography, parallel mountain ridges separated by flat valleys or basins.

Ripped away from the American Southwest, a chunk of crust containing Australia and Antarctica receded, and the ancient Pacific Ocean grew in its wake. The ocean's widening basin put northern Arizona ever farther from a tectonic plate boundary and its attendant geologic violence. Peace and quiet returned, along with erosion, which patiently leveled the Basin-and-Range-like mountains over the next 200 million years. By the dawn of Phanerozoic time 542 million years ago, the landscape was again nearly flat, with the exception of a few hills of amazingly durable Mazatzal quartzite and a few other especially hard rock units.

Since life's inception just a few hundred million years after the formation of the earth, it had been stuck in its bacterial stage, but in Phanerozoic time, life exploded in an array of complex organisms that first populated the oceans, then conquered the land. Geologists have subdivided Phanerozoic time into three eras, beginning with the Paleozoic, the era of "ancient" life. Paleozoic time was tranquil in northern Arizona. For hundreds of millions of years, a series of shallow seas encroached from the west, ultimately blanketing the region with a thick sequence of colorful sedimentary rock layers (vignettes 17 and 19). While northern Arizona experienced this interlude, plate tectonics remained active elsewhere; by 300 million years ago, near the close of Paleozoic time, a second supercontinent had formed: Pangaea. Northern Arizona lay along its west coast (vignette 13).

The collision that completed the assembly of Pangaea raised a continuous chain of mountains from the Appalachians through the Ozarks and into north Texas, where it bent northward through New Mexico and Colorado. Because it arose in the same location as the modern Rockies, the Colorado–New Mexico portion of this mighty range is called the Ancestral Rockies. In Monument Valley (vignette 6), the Grand Canyon (vignette 7), and Sedona (vignette 13), you can still see channels of ancient rivers that carried material shed from these mountains across northern Arizona. Sediment from this range inundated the region until the rivers dried up, triggering the migration of massive sand seas across the region. Dunes up to a 1,000 feet high left behind thick deposits of sand preserved in Sedona, Monument Valley, the Grand Canyon, and Canyon de Chelly National Monument (vignette 9).

Then, 251 million years ago, the Paleozoic era closed with a bang as a cataclysm of epic proportions, possibly triggered by an asteroid impact, wiped out 90 percent of species on earth (vignette 14). At the dawn of Mesozoic time, the era of "middle" life, earth's species began rebounding; during Triassic time, 251 to 200 million years ago, many strange new plants and animals burst into existence. Familiarize yourself with the Triassic landscape's ecological complexity in northern Arizona's Petrified Forest National Park, where the first dinosaurs shared the shade of tall trees with even more fearsome and bizarre creatures (vignette 15).

Triassic time also marked the beginning of the end for Pangaea, which began to split apart along boundaries of our modern continents. Pangaea's breakup inaugurated the opening of the Atlantic Ocean. The formation of new oceanic crust along the Mid-Atlantic ridge elbowed North America westward as it grew. Consequently, with a continent impinging from the east, the oceanic crust flooring the Pacific Ocean subducted under North America's west coast, forming a towering volcanic mountain chain in what is now southern Arizona. In Canyon de Chelly National Monument you can see an impressive river channel that carried sediments shed from this growing range across the region, depositing the famous Shinarump conglomerate (vignette 9)—the same rock that was crucial in sculpting Monument Valley's iconic symbols of the American West (vignette 6).

Just as life had gotten back on its feet after the greatest extinction event in history, another catastrophe—possibly another asteroid impact—hit 200 million years ago, creating a boundary between Triassic and Jurassic time (vignette 14). Although over half of earth's species went extinct, diminutive early dinosaurs in northern Arizona survived. Quickly evolving into larger predators, they tromped across the floodplains of sluggish rivers, leaving behind footprints and even a broken-off claw near Moenave (vignette 8). By 192 million years ago, another drought gripped the region, more extreme than any before, and sand dunes once again covered northern Arizona. Myriad steep, sweeping crossbeds—avalanches of sand frozen in time—adorn the thick, salmon-colored Navajo sandstone that graces much of the spectacular scenery in the American Southwest. You can see this beautiful rock in Antelope Canyon's precipitous slot (vignette 4), Navajo National Monument's archaeological ruins (vignette 5), and even the abutments of Glen Canyon Dam (vignette 3).

During succeeding Cretaceous time (146–65 million years ago), continued subduction off North America's west coast thrust up a mountain range, the Sevier-Mogollon Highlands, in Nevada and southern Arizona (vignettes 1 and 10). This range contributed thick layers of sediment across northern Arizona. At the same time, a vast inland sea to the east periodically flooded the area. The shoreline 80 million years ago was a low, swampy bog in which plant material accumulated and was buried, eventually transforming into the Black Mesa coal that today provides electricity throughout the Southwest (vignette 5).

Mesozoic time closed with a bang 65 million years ago when a killer asteroid collided with the earth, gouging out a 120-mile-wide crater in Mexico's Yucatan Peninsula and extinguishing 70 percent of all species, including the dinosaurs (vignette 14). At about the same time, after almost 700 million years of relative calm, vigorous tectonic activity returned to northern Arizona in the form of a mountain-building episode called the Laramide Orogeny, which began about 70 million years ago. Subduction off the California coast raised northern Arizona to form the modern Colorado Plateau. The Laramide Orogeny also uplifted even higher mountains around the plateau's flanks, including the Colorado and New Mexico Rocky Mountains, and rejuvenated the Mogollon Highlands in central and southern Arizona. Together with the Sevier Highlands, the Mogollon Highlands guarded the plateau's western and southern edges. Atop the Colorado Plateau, the Laramide compression reactivated ancient Precambrian faults, which slowly warped overlying sedimentary layers into large, single-sided folds called monoclines. Erosion has since carved into these great creases, exposing some of northern Arizona's most spectacular sedimentary scenery in Monument Valley (vignette 6), Canyon de Chelly National Monument (vignette 9), and Navajo National Monument (vignette 5).

The Laramide compression finally subsided around 40 million years ago, and rivers draining the eroding mountains flowed northeast across the Colorado Plateau. But events in California would soon invert the region's topography, reversing the streams' flow and eventually carving the Grand Canyon. About 30 million years ago, the subduction zone that had persisted off the continent's west coast for over 200 million years began to choke when it tried to swallow a mid-ocean ridge—a feature along which tectonic plates pull apart (vignettes 1 and 10). This event gave birth to California's San Andreas fault and ultimately caused the demise of the Sevier-Mogollon Highlands. No longer supported by compression from the subduction zone, the highlands began to spread and collapse under their own weight. This spreading was accomplished along a series of north-south trending normal faults that dropped a series of valleys down between narrow mountain ranges (vestiges of the once mightier highland), creating today's Basin and Range province. This landscape of parallel mountain ridges separated by wide basins was described by one astronaut as resembling a herd of caterpillars all heading north.

This plate reorganization triggered a massive wave of volcanic activity that began to "rock" the region about 30 million years ago. One of the later eruptions, which was much larger than any in recorded human history, engulfed the Kingman area. By tracing an 18.5-million-year-old ash flow spawned by this eruption, you can track the dramatic landscape changes in northwestern Arizona since that time (vignette 10).

The collapse of the highlands began in Nevada and California, but by 16 million years ago, northwestern Arizona was also foundering. By 10 million years ago, the stretching that formed the Basin and Range had reached central Arizona, opening up a series of northwest-angling basins in which runoff pooled to form ephemeral lakes. One of these is the modern Verde Valley. Since then, Mother Nature has carved the limestone that accumulated there into both a classic sinkhole and alcoves ideal for Puebloan villages (vignette 18).

The formation of the Basin and Range had a profound influence on the Colorado Plateau. The transition between these provinces is one of the most abrupt on the North American continent (vignette 1). The plateau's western flank, which before had ended at the toes of a mountain range, now truncated along a dramatic band of cliffs that plunged into the Basin and Range. Steep creeks began to drain this spectacular escarpment, efficiently eroding their bedrock, then carving headward, lengthening themselves. In this way, one particular stream in the Lake Mead area carved its way into the heart of the Colorado Plateau, forming the western portion of the Grand Canyon as it went. About 5 to 6 million years ago, it breached the last ridge separating it from the upper Colorado River, captured the river, and for the first time sent it coursing through the Grand Canyon to the Gulf of California (vignette 1).

As Basin and Range stretching crept eastward across northern Arizona, it thinned the crust, bringing hot mantle rocks unusually close to the surface. This extra dose of heat produced chambers of magma, which erupted in volcanic fields across northern Arizona. In the flat plateau country around Flagstaff, over six hundred different volcanoes popped up like pimples during the last 6 million years, including San Francisco Mountain, Arizona's tallest (vignette 11). One basalt flow tumbled into the gorge of the Little Colorado River, blocking its channel and forcing it to carve a new path, thus forming the state's largest waterfall (vignette 12). From another volcanic field to the northwest, a series of basalt flows poured straight into the Grand Canyon in a spectacular conflict of fire and water, damming the Colorado River multiple times (vignette 2).

Just 50,000 years ago, while the volcanoes around Flagstaff were still fuming, an asteroid one-third the size of a football field slammed into the high plains to the east, blasting out a mile-wide crater—the best-preserved impact crater on earth (vignette 14). In the most recent major geologic drama, Sunset Crater erupted a mere 900 years ago (vignette 11), chasing away some of northern Arizona's human inhabitants and promising more volcanic activity in the area's geologically near future.

Subtle changes continue to this day. Some of these are natural, such as the formation of travertine limestone deposits at numerous springs. Near Payson, energetic Pine Creek has tunneled beneath one particularly massive plug to form the largest natural travertine arch in the world (vignette 19). Other changes result from human exploitation of natural resources. Mining has noticeably altered the landscape (vignettes 5 and 17), and the construction of dams has triggered damaging and often unforeseen environmental consequences (vignette 3).

The result of this long and tumultuous geologic history is a landscape of unsurpassed beauty and geologic diversity. We hope you find this book a worthy companion as you explore northern Arizona—both its world-famous and its more secret destinations—uncovering the stories locked in stone.

A Grand Transition
PEARCE FERRY

In one of the most abrupt transitions on the North American continent, two great physiographic provinces meet in the northwestern corner of Arizona. Here at the edge of Lake Mead, you can see firsthand the Colorado Plateau's high, flat-lying mesas end in a precipice, towering over a broad, gentle valley known as Grand Wash. This valley marks the beginning of the Basin and Range province, a landscape of parallel mountain ridges separated by wide basins.

After its 278-mile journey through the Grand Canyon's soaring corridors, the Colorado River exits its secretive labyrinth and spills out into this expansive valley. Had you visited this same spot just 6 million years ago, however, the canyon and the river would have been absent. Seventeen million years ago, you would not have seen any basins or ranges either. Instead, you would have been exploring the foothills of a mighty mountain range of Andean scale that soared above the neighboring Colorado Plateau.

Even more incredible than the stunning landscape near Pearce Ferry today is its geologic history. The area's beautifully exposed rocks and grand vistas make it possible to piece together the story of wholesale landscape changes that have swept this region over the last 20 million years, razing a mighty mountain range, opening this broad valley, and ultimately carving the Grand Canyon.

High above Pearce Ferry, Grapevine Mesa (stop 1) is an ideal spot from which to observe the striking differences between the two provinces. To your right rises the escarpment of the Grand Wash Cliffs, which form the western edge of the Colorado Plateau. Visible in the cliff face are the flat-lying sedimentary-rock layers characteristic of the plateau, which stretches from here to western Colorado. To your left, parallel basins and ranges are field marks of the very different Basin and Range province, which extends westward to the edge of the Los Angeles basin.

From your perch, follow the languid waters of Lake Mead eastward to where they disappear into a large, dark canyon carved into the escarpment. You are looking at the western end of the Grand Canyon. Given the geologic importance of the Colorado River—the master stream that drains most of the American Southwest—and the fact that the Grand Canyon is one of the natural wonders of the world, geologists have expended

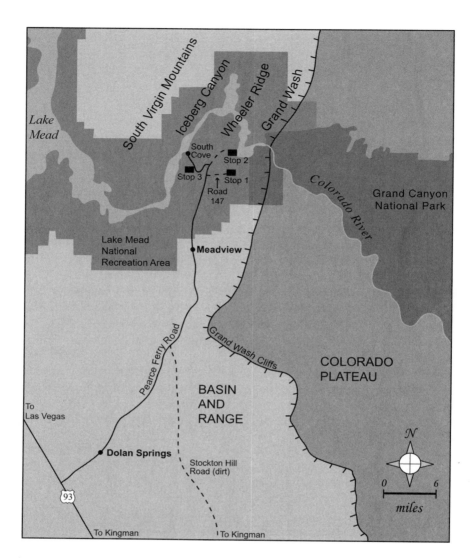

❖ GETTING THERE

Pearce Ferry is near Lake Mead in northwestern Arizona. On U.S. 93, at milepost 42, turn north onto Pearce Ferry Road (Mohave County Road 25) for 43.7 miles, through the towns of Dolan Springs and Meadview. Turn right 0.7 mile past milepost 43 onto a dirt road (Road 147) that leads to the airport. Continue 3.2 miles, staying left at each of two forks, to the north edge of Grapevine Mesa and the dirt airstrip. Check for planes, then cross the airstrip and park. Walk 30 feet to the mesa rim for an amazing vista at stop 1. To reach stop 2, return to Pearce Ferry Road and turn right. The road descends steeply off the mesa. At the signed road junction at milepost 45, continue straight 4 miles on the now gravel road to the Pearce Ferry boat ramp. To reach stop 3, go back to the junction at milepost 45. Turn right (west) onto South Cove Road. After 1.5 miles, the road passes through an impressive roadcut through Wheeler Ridge. Continue to a huge pullout on the right immediately past milepost 2. Park and cross the road to examine the roadcut.

significant effort for almost 150 years trying to understand when and how the Colorado River cut this incomparable gorge. The river's story has proven much more difficult to unravel than scientists first expected, but many aspects are becoming clear.

No single location has contributed more evidence to our growing understanding of the Grand Canyon story than Pearce Ferry, and it has become increasingly apparent that the history of the Grand Canyon is inextricably linked with the major landscape transition in front of you. But how, exactly, have geologists pieced these puzzles together? Geologists are scientific storytellers, using clues found in the rocks to weave together vivid tales of events in the earth's long history. In a setting as vast and varied as this, geologists first try to understand the big picture—the general sequence of events—before delving into details.

First, some orientation. Pearce Ferry lies 1,700 feet directly below you, at the end of the prominent dirt road. Once the site of a ferry across the Colorado River, Pearce Ferry now serves as Lake Mead's easternmost boat launch when the water level is high enough.

The Grand Wash Cliffs to your right are composed of the same rock layers exposed in the Grand Canyon's lower walls. The second rank of cliffs behind and above them, known as the Upper Grand Wash Cliffs, contain layers identical to those in the Grand Canyon's upper walls. These range from the bright red sandstone of the Supai group up to the rim-forming Kaibab formation.

To your left, west of the boat launch, you will see just beyond Pearce Ferry Road an undulating spine of steeply eastward-tilted rock known as Wheeler Ridge, bordering Grand Wash on the west. From this vantage point, because of its curvature, the spine appears to form three separate ridges. From left to right, these appear progressively farther away. The waters of the reservoir disappear into a notch in front of the easternmost ridge.

Wheeler Ridge's layers are composed of rocks of different colors and varying thicknesses. Notice the prominent band of red rock on the rightmost ridge. This consists of the Supai group, a layer that matches rocks exposed in the upper Grand Wash Cliffs. Geologists have discovered that the rocks of Wheeler Ridge once lay adjacent to those of the Grand Wash Cliffs in an extension of the Colorado Plateau. Their current separation is due to movement along the Grand Wash fault, which runs along the base of the cliffs. The vertical distance between the two matching bands of rock, known as their fault slip, is a whopping 15,000 feet—nearly 3 miles.

But how did activity along the Grand Wash fault cause the rocks in Wheeler Ridge to go from horizontal to such a near-vertical tilt? The answer lies in an understanding of the fault's geometry. Like most faults

The vista looking north from Grapevine Mesa. The Grand Wash Cliffs mark the western edge of the Colorado Plateau. Wheeler Ridge is the easternmost range in the Basin and Range. Grand Wash is the valley that separates these two features.

in the Basin and Range, the Grand Wash fault begins its journey into the earth at a steep, 60-degree angle. At deeper levels, however, the fault gradually flattens out, making it what geologists call a listric fault. When the block on one side of a listric fault drops down along this curving trace, it can't help but rotate, just like accordion door panels moving along a curved track.

Clues in the area can help us figure out when the Grand Wash fault was active. Look below you at the low, corrugated-looking hills surrounding Pearce Ferry. These hills consist of sedimentary rocks belonging to the Muddy Creek formation, the rock unit that provides the most important clues for our story. Dating of volcanic ash layers in the Muddy Creek indicates that it was deposited very recently, between 15 and 6 million years ago. In contrast, the layers composing the Grand Wash Cliffs were laid down between 525 and 270 million years ago.

Now look back toward the middle section of Wheeler Ridge, where Pearce Ferry Road executes a prominent S-turn. Between the road and the barren, steeply tilted rocks of the ridge lies a sparsely vegetated hill as broad and rounded as a whale's back, whose layers also tilt east—but not nearly as steeply, only about 20 degrees. This whaleback hill consists of the Muddy Creek formation. The difference in tilt causes its layers to truncate—to butt up at an angle—against the older rocks in Wheeler Ridge's westernmost section.

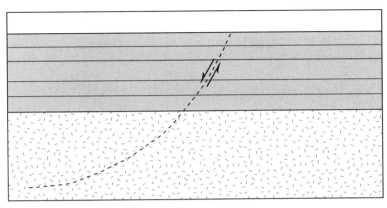

a. before fault movement

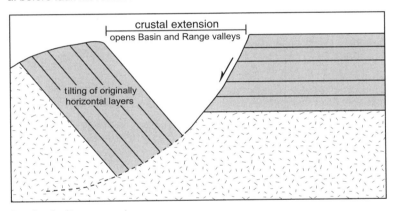

b. after fault movement

Wheeler Ridge's once horizontal layers were dramatically rotated along a listric normal fault.

Because sedimentary rock layers are always deposited nearly horizontally, this angular meeting of different-aged rocks, known as an angular unconformity, provides important information regarding the area's tectonic history. In this case, the unconformity tells us that movement on the Grand Wash fault must have begun prior to 15 million years ago, the age of the oldest Muddy Creek beds, because the ridge's older rocks must already have been substantially tilted before any Muddy Creek sediment was deposited. Furthermore, because the Muddy Creek beds are also tilted, we know that motion on the fault continued more recently than 15 million years ago, simultaneously tipping both the ridge and the Muddy Creek formation by an additional 20 degrees.

Progressive tilting along the Grand Wash fault has created an angular unconformity between the steeply tilted rocks of Wheeler Ridge and the more gently tilted rocks of the Muddy Creek formation.

We can ascertain the timing of this fault-induced tilting even more accurately with a couple of additional observations. The rocks you are standing on at Grapevine Mesa consist of a prickly, gray rock known as the Hualapai limestone, which, as we will further explore, was deposited between 11 and 6 million years ago. As you can see by looking at the mesa rim, the limestone is barely tilted. This observation tells us that fault movement must have ended by 11 million years ago; otherwise, this limestone would also be strongly skewed.

When did fault movement begin? In the nearby White Hills, geologists have found that all rocks older than 17 million years possess the maximum tilts, whereas 16-million-year-old rocks are less steeply inclined. Combining all this information, we can tell that the Grand Wash fault, which created the abrupt boundary between the Colorado Plateau and the Basin and Range here, began to move about 16.5 million years ago and ceased activity by 11 million years ago.

One last noteworthy feature is visible from the mesa top. West of Wheeler Ridge is a narrow valley, behind which another mountain range

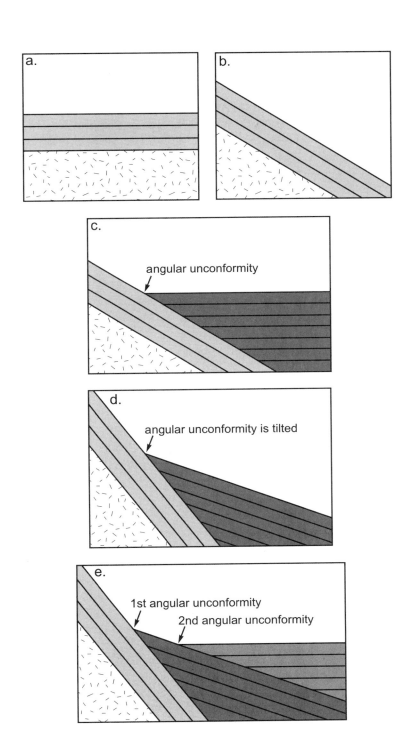

Progressive movement along a fault tilts rock layers, forming an angular unconformity. Continued movement creates additional angular unconformities.

Grapevine Mesa consists of horizontal layers of Hualapai limestone, deposited 6 million years ago. The horizontality of these beds indicates that no tilting along the Grand Wash fault has occurred since their deposition. The South Virgin Mountains are visible to the right of the limestone.

rises. This valley, whose floor is blocked from view, is Iceberg Canyon, and the mountain range beyond it is the South Virgin range. These mountains are constructed of ancient granite and metamorphic rocks dated at 1,400 to 1,700 million years old. From clues similar to those you've observed here, geologists have learned that Iceberg Canyon was opened up by movement along a second normal fault, called the Wheeler fault, which began its period of activity after the Grand Wash fault ceased to move. Unlike the horizontal Hualapai limestone at Grapevine Mesa, outcrops of the same limestone west of the Wheeler fault are tilted down to the east, indicating that the Wheeler fault accomplished much of its movement after the completion of limestone deposition 6 million years ago. If it is not too hazy, you may also be able to make out Gold Butte, the highest peak in the South Virgin range. This peak is made of a distinctive granite body known as the Gold Butte granite, which will play a significant role later in our story.

From this overlook, you have been able to piece together a broad sketch of the area's geologic history, which goes something like this: About 17 million years ago, the area now occupied by Wheeler Ridge formed a westward continuation of the Colorado Plateau. Between about 16.5 and 11 million years ago, movement on the Grand Wash fault stretched the area out like an accordion and down-dropped Grand Wash, creating the abrupt transition between the Basin and Range and the Colorado Plateau that exists today. Then, beginning about 6 million years ago, the Basin and Range was further stretched when the Wheeler fault came to life and opened up Iceberg Canyon to the west.

Clearly this region has experienced major upheavals in the last 17 million years. The more geologists learn about the area, the more they realize that the birth of the Basin and Range province must have had profound significance for the history of the Colorado River. Today this river flows out of the Grand Canyon into Grand Wash; but where did it go 17 million years ago, when the Grand Wash valley didn't even exist? When was the canyon carved? Even today, geologists have only partial answers to these questions, based in large part on the evidence preserved in the Muddy Creek formation, which we will now examine in detail.

The Muddy Creek formation can be subdivided into three distinct units based on differences in the rock types it contains. The uppermost

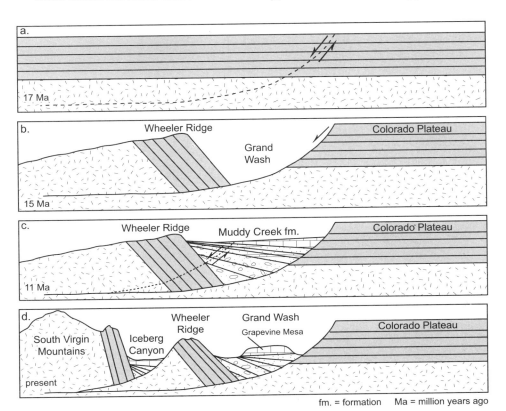

Development of the Pearce Ferry area over the last 17 million years

unit, and hence the youngest, is the Hualapai limestone under your feet. Below this limestone, the middle unit consists of interbedded sandstone and siltstone, with a few volcanic ash and gypsum layers thrown into the mix. We'll look at this material at stop 2. Finally, the lowest and hence oldest rocks are known as "fanglomerates," mixes of exceptionally large, angular rock fragments embedded in finer-grained materials. Stop 3 provides a great opportunity to examine these.

Each Muddy Creek unit can tell us something about the location of the Colorado River when that unit was deposited and, ultimately, about when and how the Grand Canyon was carved. From Grapevine Mesa we can start at the top and work our way down. The Hualapai limestone is a typical limestone: hard, gray, and rough. Most limestones form either in the ocean or in a lake, and fossils usually provide the most definitive

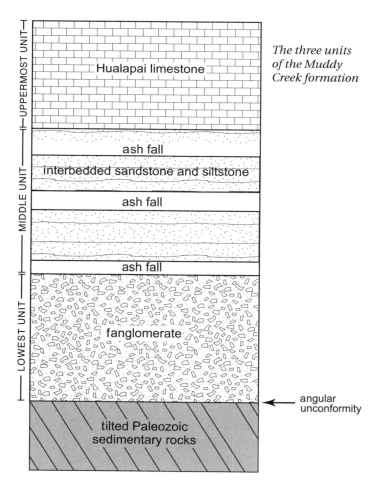

The three units of the Muddy Creek formation

environmental clues. The Hualapai limestone has not served up a rich fossil record, but what it has yielded suggests that it was deposited in a lake, and other lines of evidence support this conclusion. Could this lake have been fed by the Colorado River? There are a few problems with this idea. Before Glen Canyon Dam began trapping Colorado River sediments in the still waters of Lake Powell (for more on that story, see vignette 3), the Colorado was one of the muddiest rivers on the continent, moving tens of millions of tons of sand and mud downstream every year. If this lake had been fed by the Colorado River, where is all this mud? Furthermore, the Colorado in flood can move pebbles and cobbles downstream; wouldn't we also expect to see some of these larger stones? No such mud or gravel deposits have been found in the area, suggesting that this ancient lake was not fed by the mighty Colorado River.

To inspect the Muddy Creek's middle unit, proceed to stop 2. All of the low hills surrounding Pearce Ferry consist of this rock, but a particularly good place to examine it is the small peninsula to the right of the boat ramp. Here, just above the high-water line, a distinctive white band

An ash layer (light-colored) overlies mudstones and sandstones in this outcrop of the middle unit of the Muddy Creek formation. The horizontal layers of the Grand Wash Cliffs loom behind.

10 inches thick rests on top of brown layers. This white layer is volcanic ash. The uniformly small size of its particles indicates that it didn't come roaring down the flanks of some nearby volcano, but rather fell like snow upon the landscape, blown here on the prevailing winds from a distant volcano. The layers below the ash consist mostly of brown silt and fine sand, with a number of pebble beds and a few light-colored evaporite layers, which form when mineral-laden water evaporates, sandwiched in between.

Could the Colorado River have delivered these sediments? The clues to the answer are subtle. If you look more closely, you will see thin beds of brown sand mixed in with the white ash layer. The presence of this contaminating material indicates that, as the ash particles drifted toward the ground, they were gently stirred and shifted around, incorporating some of the underlying sand. The stirring couldn't have been too vigorous, though, or the ash layer would have lost all coherence and been completely mixed in. So the environment in which this ash settled was dominated by gentle currents, such as you would find in a shallow lake or a sluggish stream feeding it.

Can we figure out what type of environment it was? Take a short walk to the top of any of the low hills nearby to get a good, close-up look at the sandstone, siltstone, and evaporite layers en route. Try to envision a modern environment that would produce such a pattern of layering. These sediments are identical to what you would find if you dug a trench through the edge of a modern Basin and Range playa, such as Death Valley. A playa is the low point of a desert valley with no drainage outlet. Water accumulates seasonally in the valley bottom but readily evaporates, leaving its salts behind. When the playa contains water, silt settles out, as do coarser sand and pebbles when they wash down into the valley during periodic floods. Therefore, as with the Hualapai limestone, this middle unit of the Muddy Creek formation displays the characteristics of having been deposited in a closed valley that received its material from local sources, not from a big river like the Colorado. But the evidence isn't ironclad. Let's look for more definitive clues in the lowest unit, the fanglomerate layer in Iceberg Canyon valley.

The roadcut at stop 3, in Iceberg Canyon valley, is a chaotic jumble of material of all sizes. One boulder is over 15 feet across. Like many of the chunks in this deposit, this boulder has sharp, angular edges. Because the corners of boulders quickly round off when transported by rivers, the angular shapes of the chunks tell us that they didn't move very far from their source. The sheer size of some of these boulders reinforces this conclusion; even powerful rivers have trouble moving such big rocks very far. Fanglomerates like this typically collect on steep alluvial fans situated at

Fanglomerate, the lowermost unit of the Muddy Creek formation. The 15-foot boulder pictured here, on the left side of the cut, consists of Gold Butte granite.

the base of cliffs. It appears these chunks did not travel in a river like the Colorado.

In addition, all of the chunks in this fanglomerate came from rock units in the mountains flanking Grand Wash; none came from far to the east, as we would expect if the Colorado River had carried them here. Most telling are the many granite chunks, including the 15-footer. The big, pink feldspar crystals and other characteristics unambiguously identify them as Gold Butte granite, which only occurs west of here, in the South Virgin range.

These observations make our growing suspicion—that the Colorado River didn't deposit the Muddy Creek formation—indisputable. Not only did the Colorado River not lay down these sediments, but the presence of the granite chunks shows that local drainage 16.5 to 6 million years ago flowed west to east—the opposite direction of the modern Colorado River. However, the discovery of 4.4-million-year-old conglomerates full of rounded river cobbles a short distance west of here indicates that the vigorous Colorado River had indeed begun to flow here by that time.

So the Colorado River found its way into Grand Wash from the adjacent western Grand Canyon sometime between 6 and 4.4 million years ago. The western Grand Canyon is therefore much younger than geologists previously suspected. Now that we know when the Grand Canyon was cut, the question becomes, exactly *how* was it carved? Geologists disagree about this issue, but given the evidence at hand, the most plausible and currently most accepted hypothesis is, in some ways, the most startling. It is also integrally linked to the development of the Basin and Range. The story goes like this.

Prior to 17 million years ago, the Andean-scale mountain range mentioned before stood to the west, where the Basin and Range lies today. The rivers of this region flowed down onto the Colorado Plateau from southwest to northeast. Beginning 16.5 million years ago, movement on the Grand Wash fault opened the Grand Wash valley, cutting off this drainage. Rotation along the listric fault tilted and uplifted the rocks of Wheeler Ridge, which at that time also included the entire mass of the South Virgin range. The Colorado Plateau, which earlier had been lower than the land to the west of it, now stood several thousand feet higher than Grand Wash. A small, steep stream began to drain off the new escarpment, bringing with it sediments that accumulated in the valley. The newly formed Wheeler Ridge mountain range was even taller, so it contributed the bulk of sediments. The growing pile became the fanglomerate of the Muddy Creek formation.

For 3 million years the Wheeler Ridge mountains eroded at a very high rate, becoming lower while the adjacent Grand Wash filled with sediment. Meanwhile, the small, steep stream draining the Grand Wash Cliffs continued to lengthen, eating its way eastward into the Colorado Plateau. All rivers are steepest at their headwaters and erode most efficiently where they are the steepest. Consequently, through a process known as headward erosion, rivers lengthen themselves over time by eroding farther and farther back into their headwater terrain.

As movement along the Grand Wash fault ground to a halt and the Wheeler Ridge range became more subdued, accumulation of fanglomerate gave way to the playa sand, mud, and gypsum evaporites that you observed at Pearce Ferry. Because Grand Wash had no outlet, all the water and sediment that flowed into it ended up accumulating there, forming the middle Muddy Creek unit. By 11 million years ago, a permanent lake formed on the Grand Wash valley floor, and the Hualapai limestone accumulated in it.

Through continued headward erosion, the small stream referred to as the Hualapai drainage, which had cut into the Grand Wash Cliffs, had grown much longer. Although it still did not carry much water, its steepness allowed it to carve out a respectable gorge as it went, forming what we now call the western Grand Canyon. A growing body of evidence suggests that this carving was assisted by several catastrophic floods, perhaps triggered by the sudden drainage of large lakes in northeastern

Arizona. By 6 million years ago, the lake in which the Hualapai limestone had been accumulating dried up, and movement along the new Wheeler fault disrupted what was left of the Wheeler Ridge mountain range. The movement opened up the valley of Iceberg Canyon and shifted the South Virgin range west to its current location. With Wheeler Ridge eroded and now dismembered, water and sediment spilled out of Grand Wash for the first time, sending a river south to the newly formed Gulf of California.

The Hualapai drainage into Grand Wash from the east thus merged with the drainage out of the wash to the west. This network would soon become the path of the lower Colorado River, but another dramatic event had to occur first. This event was likely what geologists call stream piracy. Because streams are constantly growing headward, it is inevitable that from time to time, one will encroach on another. When the steep "pirate" stream has carved away the last bit of land separating the two, it "beheads" its conquest and spirits its waters away.

Most geologists believe that the ancestral upper Colorado River was captured in just such a manner in the eastern Grand Canyon sometime between 6 and 4.4 million years ago. When the swashbuckling Hualapai pirate stream sliced through the last ridge separating it from the Colorado,

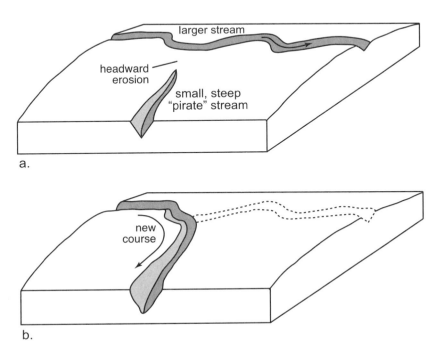

The process of stream piracy

it instantly changed from a trickle to a raging river. The dramatically increased water flow, combined with the steep gradient, set this new river to carving its canyon deeper and deeper at a very rapid rate, ultimately forming the spectacular western Grand Canyon.

When the river spilled into the Grand Wash valley, it cut down through the sediments of the Muddy Creek formation, creating the escarpment that separates the top of Grapevine Mesa from Pearce Ferry. And, by so doing, it exposed the very rock unit that holds the clues to the river's history, here near the remarkable physiographic transition that first set that history in motion.

A Conflict of Fire and Water
TOROWEAP OVERLOOK

During his remarkable first descent of the Colorado River in 1869, geologist and explorer John Wesley Powell marveled at the cascades of once molten lava on the stunningly vertical walls of the western Grand Canyon. "What a conflict of water and fire there must have been here!" he exclaimed in his journal. "Just imagine a river of molten rock running down a river of melted snow. What a seething and boiling of the waters; what clouds of steam rolled into the heavens!" Today you don't need to spend months rafting down the Colorado River to see the western canyon's narrow, steep gorge and its lava cascades. Visit the North Rim's breathtaking Toroweap Overlook instead, one of the Grand Canyon's best-kept secrets.

At Toroweap, cascades of dark basalt lava break the precipitous walls of the Grand Canyon's inner gorge to the right of the river. Note the small cinder cone on the southern (left) *canyon rim just above the sun-shade line.*

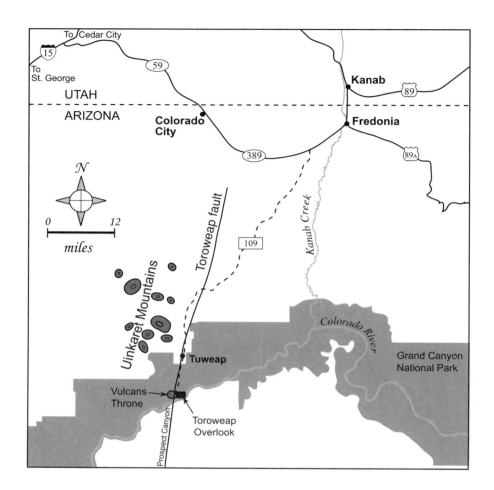

❖ GETTING THERE

Remote Toroweap Overlook, stop 1, is accessed by a 60-mile dirt road that is perfectly passable in passenger cars in good weather. No services are available in this northwestern part of Grand Canyon National Park, so make sure you are prepared (see preface) before setting off. From Fredonia, travel west on Arizona 389 for 7 miles, then left on BLM 109 for 54 miles to the Grand Canyon National Park boundary and Tuweap Ranger Station (not always staffed). At the fork 3.5 miles past the ranger station, go left. After 2 miles, at a second fork, go right (south) another 0.8 mile to the overlook parking area. The best vistas are from an obvious promontory 200 yards farther west. Exercise extreme caution along the rim. To reach stop 2, return to the fork 3.5 miles south of the ranger station and take the unmarked dirt road to the left. This dead-ends at the canyon rim after 2.4 miles. Low-clearance vehicles may have trouble with thick dust 0.5 mile in; if so, consider continuing on foot, about 45 minutes one way.

After parking at stop 1, approach the overlook for your first glimpse into the abyss. The dramatic view is both vaguely familiar and radically different from the typical postcard scenes of the eastern Grand Canyon. The canyon's geology is by and large the same in both its eastern and western portions, consisting of a thick cake of brightly tinted, flat-lying layers sliced by a deep river gorge. However, two characteristics of this western portion of the canyon do not square with the standard image: the gorge's startling sheerness and the abundance of volcanoes, whose black outpourings so impressed John Wesley Powell. In this vignette, we will examine these dramatic differences and what they reveal.

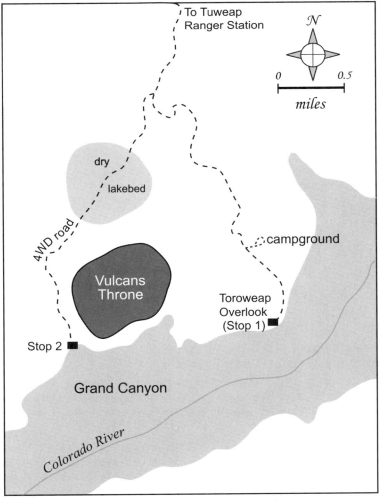

Detail map of the Toroweap Overlook area

The walls of the eastern Grand Canyon drop from rim to river in a series of giant stair steps, with risers formed of strong limestone or sandstone and treads consisting of weaker mudstone.

In contrast to the stair steps of the eastern Grand Canyon, the gorge of the western Grand Canyon near Toroweap plunges 3,000 feet straight to the river.

In the eastern Grand Canyon, most visitors are awestruck by the vastness and complexity of the 10-mile-wide hole into which they are gazing. The landscape steps down to the Colorado River in a series of cliffs resembling a giant's staircase. From the rim, you can catch only fleeting glimpses of the river below. In stark contrast, at Toroweap the river lies in plain view at the bottom of a vertical, 3,000-foot-deep chasm, nearly as deep as it is wide—a most unusual attribute for a large canyon.

The striking differences between the eastern and western Grand Canyon boil down to one deceptively simple material: the soft sedimentary rock known as mudstone. Sheer canyons like the one at Toroweap form only when a river is cutting down into a mass of strong, resistant rock. Thick sequences of sedimentary rocks like the one you see here typically alternate between sandstones and limestones—materials strong enough to support steep canyon walls—and mudstones—which are easily eroded and usually form gently sloping benches.

As the Colorado River carved the eastern Grand Canyon, it first encountered resistant limestone, which supported a narrow canyon. But as the river continued to eat its way down, it soon met a bed of mudstone. The river devoured these soft rocks, undercutting the steep limestone cliffs above and causing them to collapse along weak joints (cracks in the rock) in a series of landslides. This process, known as scarp retreat, widened the canyon. Cutting down, the Colorado alternately encountered hard and soft sedimentary layers. At every resistant layer, the river carved a narrow canyon immediately above its banks, and at every soft layer, another wave of scarp retreat began. The result is the eastern canyon's extremely wide stair-stepped appearance.

The precipitous narrowness of the western Grand Canyon indicates that scarp retreat has played only a minor role here. From rim to river, the sheer walls in front of you consist of layers of sandstone and limestone, with almost no mudstone. Why is there so little mudstone at Toroweap? First, the same rock formation can vary compositionally from place to place. Due to gradual changes in sedimentary environments in the region 280 to 270 million years ago (more about these in vignette 7), layers of rock that consist of mudstone in the eastern canyon are instead composed of more resistant rock here. Second, the Bright Angel shale, a thick mudstone layer that forms the most striking bench in the eastern canyon, is only just now being excavated at Toroweap because the western canyon was not lifted as high during the area's most recent episode of mountain building (more on the Colorado Plateau uplift in vignette 9).

Finally, the western canyon's only significant layer of mudstone, the brick-red Hermit formation, is nine times thicker than in the eastern canyon—for example, near Grand Canyon Village. This difference has

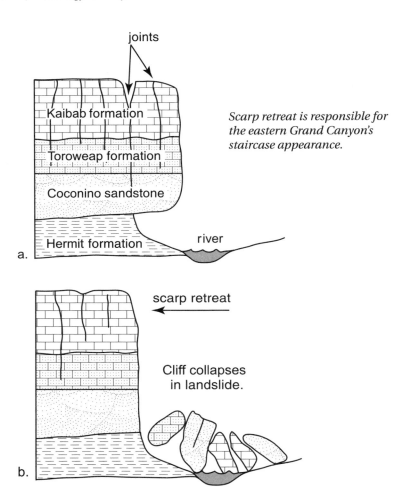

Scarp retreat is responsible for the eastern Grand Canyon's staircase appearance.

led to much more extensive retreat of Toroweap's uppermost walls—up to a couple miles from the rim—than occurs farther east. Look north at the cliff rising 1,500 feet to a promontory above you. This cliff contains the same rock layers that form the upper canyon walls near Grand Canyon Village, but here they have retreated so far that they no longer appear to be part of the canyon. The road you took to the overlook passed right through them where they were not so steep. Since your current vantage point lies below the Hermit mudstone, it doesn't form a bench partway down the cliffs to disrupt the gorge's startling sheerness.

Ultimately, then, Toroweap's vertical cliffs owe their existence to ancient events before and during the uplift of the Colorado Plateau. The area's volcanic features, however, derive from much more recent tectonic events.

View north from Toroweap Overlook. The upper cliffs of Toroweap and Kaibab limestone have retreated far from the river due to scarp retreat triggered by erosion of the soft Hermit mudstone, which forms the sloping apron below the cliffs.

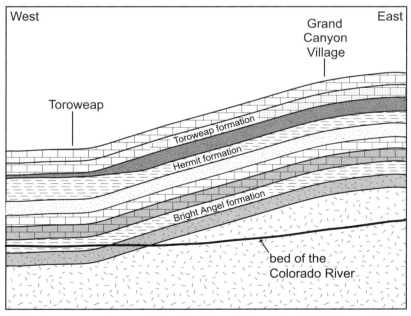

Tectonic processes have uplifted the Grand Canyon's colorful layers higher in the east, forcing the Colorado River to cut more deeply through the layer cake there, revealing more soft mudstone.

From your perch high above the river, look westward (right) down the river gorge to where, in several places, steep cascades of black basalt break the red and gray of the canyon walls. These are some of the lava cascades described by John Wesley Powell. During the last 2 million years, volcanoes have repeatedly erupted on both rims of this gorge. The nearest petrified cascade tumbles into the canyon below Vulcans Throne, the largest and most prominent cone-shaped mound visible on the canyon rim, about 1 mile downstream from the overlook on the same side of the river. See if you can pick out the rough hiking route that zigzags down this lava cascade to the river at Lava Falls rapid, the biggest rapid on the river, which foams below. The trail is grueling, suitable only for very experienced hikers. Without that layer of lava, the wall would be too sheer to hike at all.

Had you arrived here just a smidge earlier in geologic time—say, 140,000 years ago—you would have witnessed molten lava pouring into the canyon. Through painstaking field work, geologists have found that, within the last 640,000 years, lava has poured into the Grand Canyon at this very spot no fewer than thirteen separate times. Each flow filled the narrow gorge from wall to wall, forming a plug that dammed the flow of the Colorado River. In other, wider parts of the canyon with more mudstone, the lava would not have been able to stretch from one wall to the other to form these high dams.

Different lava dams formed at different rates, with some constructed in a matter of weeks and others growing episodically over years. Nobody knows for sure if any were monolithic enough to hold back the entire flow of the Colorado River, or if cracks and weaknesses within the structures offered the river an escape valve. If, as many geologists believe, some of the dams were tight enough not to leak, they would have impounded the Colorado's waters in reservoirs analogous to the modern ones humans have created on the river. But some of these lava dams were much larger. The titanic Prospect dam, which formed 640,000 years ago, probably created a reservoir well over 400 miles long (versus Lake Powell's 186 miles), inundating the entire Grand Canyon and modern Lake Powell and continuing upstream all the way to Moab, Utah. Not only would this lake have reached almost all the way up to the ledge where you are standing, it also would have lapped at the base of the Redwall limestone cliff near Grand Canyon Village, filling the canyon there half full.

Proceed to stop 2, Vulcans Throne. Vulcans Throne is a cinder cone—a type of volcano with smooth, steep sides and a circular depression at the top where the vent was located. The cinders are solidified fragments of lava that cooled high in the air, thrown there by the sputtering volcano 74,000 years ago. Descending, they amassed around the vent to form this

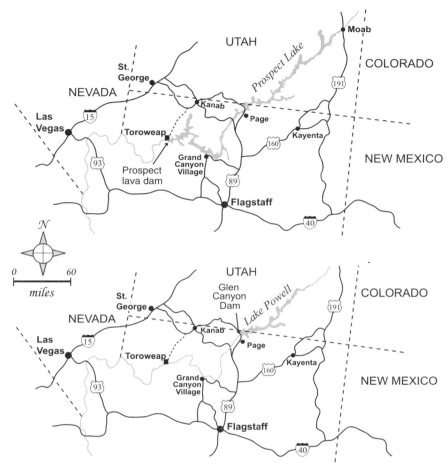

A lake impounded by the Prospect lava dam would have been three times longer and three times deeper than modern Lake Powell.

mound. The loose rocks that litter this area are cinders ejected from that eruption. Look for another cinder cone across the river, perched on the edge of Prospect Canyon.

Gazing over the rim at stop 2, you can still see the eroded remnants of several lava dams plastered on the far side of the gorge (you can also see remnants in the photo on page 28, bottom—the dark material left of the river at photo center). Superimposed upon each other, they are almost impossible to tell apart visually, but geologists have distinguished remnants from four different dams in this spot. Across the river, look for dark rock plastered against the canyon wall all the way up to the broad, reddish layer or platform of rock on which you are standing, here at the rim of the inner gorge. That dark rock is a scrap of the huge Prospect dam. From here you can appreciate the incredible height at which it towered above the river.

If lava dam after lava dam rose in the same location, what happened to earlier versions? The force of the Colorado River eventually dismantled each one, swept it away before the next dam took its place. The exact processes involved in breaching a lava dam would depend on the dam's strength. If the dam were largely watertight, a vast reservoir would pond behind it until the water overtopped it. Just downstream of where you're standing, water would have poured over the Prospect dam's face, creating a spectacular waterfall fourteen times higher than Niagara Falls.

Such a waterfall would begin the work of destroying the dam's foundations, but because the sediment the river carried would settle behind the dam, the water spilling over would be clear, and the rate of erosion modest. The really strong dams would not fail until their reservoirs filled completely with sediment. Then, mud, sand, and boulders transported by the river would tumble over the falls and crash into the base of the dam with enormous force. No structure could survive such abrasive action for long; geologists have estimated that this process of dam obliteration would take a mere 20,000 years.

A leaky dam, however, would fail sooner, triggering yet another spectacular event. Dams that allow some but not all of the river's flow to pass can still impound large reservoirs behind them. However, water that manages to squirm its way through the dam gradually expands any structural weaknesses, until one day the dam loses all strength and collapses catastrophically. Such a failure sends a wall of water rushing downstream at tremendous depth and speed. These floods would be larger than those associated with tighter dams, which have reservoirs filled with mud, not water, by the time they fail.

Not long after Prospect dam was swept away, new eruptions created the Ponderosa dam, then the Lava Butte dam. These were followed about 560,000 years ago by the Toroweap dam, whose reservoir stretched 180 miles upstream to Lees Ferry (vignette 3), the put-in point for Grand Canyon rafting trips. Over time, the lava flows and resulting dams became smaller. The most recent one, just 140,000 years ago, rose a mere 226 feet. The presence of younger volcanic features on the rims of this gorge, including Vulcans Throne, suggests that more dams could appear in the geologically near future.

Clearly, recurring volcanic outpourings have endowed the Toroweap area with unique scenic and geologic characteristics. But why is so much volcanism focused here? It is because the Colorado Plateau and the

Grand Canyon are being ripped apart, and Toroweap is ground zero for the carnage.

After the Laramide Orogeny, tectonic peace and quiet reigned here for 35 million years. Beginning about 5 million years ago, however, in a new dynamic, plate tectonics began to tear down the plateau it had worked so hard to build up. The birth of California's San Andreas fault had repercussions even here, far to the east, where it has triggered the formation of the Basin and Range province, characterized by stretching of the earth's crust. This extension has fractured the area from the edge of the Colorado Plateau to the mountains ringing the Los Angeles basin into a series of long, narrow mountain ranges separated by deep valleys—a topography reminiscent of the pleats on a Chinese fan. (Read more about the Basin and Range in vignettes 1 and 10.)

In the Basin and Range, each mountain range is bounded by a normal fault, a break in the earth's crust along which the side forming the valley slides down relative to the side containing the mountain range. Over time, the area of normal faulting has expanded to both the east and west, progressively enlarging the Basin and Range. About 5 million years ago these faults began to invade the part of the Colorado Plateau hosting the western Grand Canyon. They haven't been active long enough to destroy the flat, high plateau's topography, but in just a few million years, this area will be transformed into a lower-elevation mountain and valley topography typical of the Basin and Range, similar to what you see around Las Vegas.

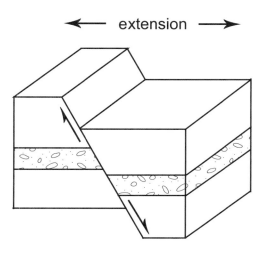

In a normal fault, the upper block of rock (the block that rests on top of the angling fault plane) slides down relative to the block below.

The two largest normal faults invading the western Colorado Plateau are the Hurricane and Toroweap faults. Where you stand, the Toroweap fault, named for this area, runs directly beneath your feet. Because faults grind and weaken rocks, canyons often form along them. Across the river, Prospect Canyon runs right along Toroweap fault, as does Toroweap Valley, the long, linear depression you descended to arrive here.

From where you stand, you can determine how much vertical motion has occurred along the Toroweap fault. The key is to pay attention to the elevation of the rim of the Inner Gorge, called the Esplanade platform, which occurs directly above the distinctive, bright red Esplanade sandstone rock layer. Looking southwest (right), toward Prospect Canyon's

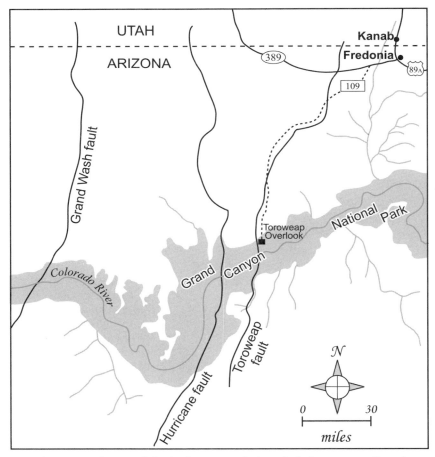

A swarm of normal faults has recently invaded the western Grand Canyon, providing conduits for magma to reach the surface.

west side, you can see that the canyon rim there is considerably lower than on the canyon's east side (see also the photo on page 25). The canyon rim developed at the boundary between the easily eroded Hermit formation mudstone above and more resistant red Esplanade sandstone below. If you examine a topographic map of the area, you would find that the fault's western side has dropped 700 feet compared to the east. Clearly there has been a substantial amount of motion along this fault since it rumbled into life 5 million years ago.

How is this action related to volcanism? When normal faults such as the Toroweap pull the crust apart, the crust thins in the process. This allows extremely hot material from the mantle to well up to unusually shallow depths. The injection of so much hot material so close to the surface triggers, not surprisingly, a lot of volcanic activity. Basaltic magma works its way upward along any weakness it can find, bursting forth at the surface to form a field of cinder cones and lava flows.

In the Toroweap area, a series of basaltic outpourings began to cover the landscape just as the Toroweap fault began to stir, and they have continued ever since. The youngest volcano erupted a mere 1,000 years ago, and future fireworks are likely. The Uinkaret Mountains, whose pine-clad slopes you can see to the northwest, are a chain of the largest volcanoes to erupt in the area.

In the Toroweap area, reminders of the earth's dynamism are omnipresent, from fresh cinder cones to the fault offset of the Esplanade platform. The images you have just entertained in your mind's eye—Powell's mighty conflict of volcanic fire meeting water, as well as giant waterfalls, collapsing lava dams, and powerful recurring floods—illustrate how profoundly a landscape can change in—geologically speaking—the blink of an eye.

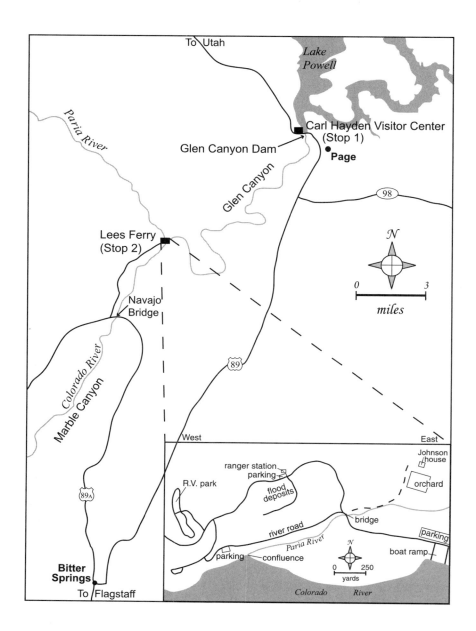

❖ GETTING THERE

Stop 1 is Glen Canyon Dam's Carl Hayden Visitor Center, located along U.S. 89 on the west side of the Colorado River 2 miles northwest of Page. Stop 2 is Lees Ferry. Proceed south from the dam on U.S. 89 approximately 25 miles to the junction with U.S. 89A. Turn right on 89A and head north 14 miles to Navajo Bridge. Cross the bridge and turn right on Lees Ferry Road. Continue 4.2 miles to a right turn (just past the left turn into Lees Ferry campground and R.V. Park) onto the river road. Proceed another 0.25 mile to a pullout on the right side of the road above the confluence of the Colorado and Paria Rivers. Stop 2 is actually a collection of sites we'll explore right around Lees Ferry.

3
Flood of Controversy
THE COLORADO RIVER AND GLEN CANYON DAM

Glen Canyon Dam is truly an engineering marvel. Rising 710 feet above bedrock and composed of enough concrete to pave a four-lane highway from Phoenix to Chicago, this monumental arch impounds Lake Powell, the country's second-largest reservoir after Lake Mead. Since the 1960s, Glen Canyon Dam, along with the slightly larger Hoover Dam downstream, has played an unsurpassed role in opening the southwestern United States to rapid settlement. Over 30 million farmers and urban residents now depend on irrigation and drinking water from the Colorado River and electricity and flood control from the more than forty dams scattered throughout the river's drainage.

There is, however, a price for these benefits. Glen Canyon Dam and its cousins have wreaked environmental havoc downstream. Due to its massive size, its situation as the first major plug on the Colorado's main stem, and its location just 15 river miles above the beginning of the Grand Canyon, Glen Canyon Dam has become one of the Southwest's most controversial environmental issues.

As the Colorado River wends its way through the stunning layer cake of soft sedimentary strata on the Colorado Plateau, it picks up and hauls tremendous quantities of sediment. This used to leave its waters so flamboyantly muddy that the Spanish explorers who stumbled across it dubbed it el Rio Colorado, "the red-colored river." Now, because Glen Canyon Dam traps the sediment coming from upstream, the Colorado River below the dam runs a clear and dazzling blue-green—incongruous in this parched and dusty setting. While gorgeous, these crystalline waters signal drastic environmental changes with serious implications. We'll explore some of those in this vignette, but first let's get acquainted with the dam.

At stop 1, park and head straight for Glen Canyon Bridge. In the pedestrian lanes, you can stroll 700 feet above the river and take in the superb views of the dam, Lake Powell, and, downstream, the lowermost, undammed portion of tranquil Glen Canyon. From this lofty perch, it is difficult to appreciate just how large Glen Canyon Dam really is. The structure is a huge wedge that tapers from 300 feet thick at the foundation's base to just 25 feet thick at its crest. On the upriver side of the bridge, below the dam's soaring white walls, is a squat, rectangular building. This is the dam's $200 million hydroelectric power plant. To give

Impressive Glen Canyon Dam impounds Lake Powell, the country's second largest reservoir.

you a sense of scale, the flat, grassy area between the power plant and the dam's base is the size of a football field. Even harder to grasp are the dimensions of Lake Powell. When full, the reservoir stretches 186 miles and covers 161,000 acres—a quarter of the size of Rhode Island. Inside the visitor center, an impressive model of the reservoir helps convey a sense of this huge scale.

After you marvel at the dam, carefully cross the highway to gaze down into the depths of Glen Canyon. As you can see, this is an extremely narrow and deep gorge slicing through sheer sandstone cliffs—to all appearances a great location for a very large dam, given that a huge volume of water can be impounded with a relatively small amount of concrete. Ironically, the driving force behind the construction of Glen Canyon Dam at this specific location was less physical than it was political. No one would have searched for a dam site in what was then one of the remotest areas of the country had it not been for an agreement reached back in 1922, when the waters of the Colorado River were apportioned like slivers of pie amongst seven ravenous states. The states were geopolitically grouped into an Upper Basin (Colorado, Wyoming, Utah, and New Mexico) and a Lower Basin (Arizona, Nevada, and California).

Map of the Colorado River basin showing its tributaries and major dams. Glen Canyon Dam is near the boundary between the Upper and Lower Basins.

Lees Ferry, which we will visit at stop 2, was selected as the dividing point between these two basins. To meet their water delivery obligations to the Lower Basin, Upper Basin states needed a means to store water, and congressional legislation in 1956 paved the way for building the requisite dam near Lees Ferry. The engineers had only to travel 15 miles upstream to find this site.

The beautiful red bedrock you see around the dam, known as the Navajo sandstone, was deposited in a vast desert here between 192 and 178 million years ago. As you return to the visitor center, look closely at the outcrops behind the building. You will notice that the rock is composed almost entirely of shiny, very hard quartz grains. If you examine it with a hand lens or magnifying glass, you will see that the grains are all about the same size, and all are worn smooth, without edges. Quartz sandstones like the Navajo are thought to be the leftovers of multiple episodes of uplift and erosion. Each erosion event dissolves away some of the less resistant mineral components, and abrasion through rough-and-tumble processes like stream transport or wave action slowly smoothes out the jagged edges on the remaining quartz crystals. It took a tremendous amount of scraping and scuffing to shape these grains; geologists have calculated that, to become so round, these grains had to have traveled a distance equal to twice around the globe before they settled in this former desert and lithified.

Normally, such homogeneous rock would be ideal for hosting a dam, but this sandstone, and consequently this dam site, has one flaw: the Navajo sandstone is, for rock, quite soft and porous—not the best rock for holding back nearly 9 trillion gallons of water. From the glass windows inside the visitor center, notice the dampness on the downstream rock walls, and the vivid green vegetation clinging precariously to them. These are visual clues that water is seeping around the dam; the Navajo sandstone absorbs water like a sponge and conveys it away from the reservoir.

Notice also the thousands of metal protrusions peppering the steep red walls downstream of the dam. These are rock bolts, put in place during dam construction to reinforce the soft sandstone. The need for this support became urgently apparent during the summer of 1983, when extremely high river flows nearly resulted in a catastrophic spillway failure. A combination of factors created dangerous conditions: inadequate forecasting, a desire to keep the lake level as high as possible to maximize power generation and recreational opportunities, and a wet winter followed by a series of warm rainstorms that swiftly melted the Colorado River basin's snowpack. The reservoir level rose so quickly that the power generation outlets could not keep pace. To avoid water overtopping the

Glen Canyon just below the dam. The prominent streaks on the lower left wall are from water weeping through the gorge's porous Navajo sandstone walls.

dam and destroying the power plant, the spillways and other outlets were opened full bore.

Most dams are designed to spill any excess water over the top, but because of its unique site, Glen Canyon Dam is different. Its two spillways run around the dam and discharge from huge, 41-foot-wide openings, one of which you can see across the river. The spillways are lined with reinforced concrete, but in 1983 the violent and irregular motions of the water and air surging through them ate hungrily through this lining and into the soft Navajo sandstone bedrock. The first graphic clue that this was occurring was when the normally clear water issuing from the spillways suddenly turned bright red, indicating that it was carrying with it chunks of Navajo sandstone.

The spillways came dangerously close to unleashing a catastrophic flood. The deluge subsided just in the nick of time; the reservoir peaked a mere 7 feet below the dam's crest, and the spillways, though severely damaged, miraculously held. Had they failed, the full force of Lake Powell would have been released, and the resulting surge might have toppled downstream dams in a dominolike failure of scarcely imaginable proportions. The visitor center hosts an exhibit that discusses the 1983 events.

It is usually possible to take tours of the dam, though these are discontinued periodically due to security concerns. Don't miss the tour if it is available during your visit.

Once you have explored the visitor center, proceed to stop 2. On most days of the year, the color of the waters of the Paria and Colorado Rivers is a study in contrasts. Colorado River water released through the dam's turbines is clear and blue-green because it contains almost no sediment; the particles settle out in the reservoir's slack waters. The Paria, on the other hand, is brown and murky like the predam Colorado. You can usually see a "mixing line," a sharp boundary between the Paria's milky water and the Colorado's clear water, that stretches a short distance downstream before the Paria's outflow assimilates into the voluminous Colorado.

The Colorado River used to transport a whopping 66 million tons of sediment past Lees Ferry each year. With Glen Canyon Dam in place, the only source now available to replenish the Grand Canyon's sandbars and beaches comes from tributaries that enter below it. The Paria is one of the principal sources of that sediment.

To better gauge the availability of sediments in the postdam Colorado River, geologists have developed a system known as a "sediment budget." Just as you balance income and expenditures in a checkbook to avoid overdrawing your account, a sediment budget tracks both the incoming supplies of sediment and the outgoing losses to determine if the system is in the black or the red. Due to the soft, easily eroded sedimentary rocks it traverses, the Colorado used to have a robust sediment income. Now, at least through the Grand Canyon, the river is nearly bankrupt.

Today, an estimated 70 percent of the river's sediment income comes from the Paria and another tributary, the Little Colorado River. The Little Colorado enters the main stem 60 river miles below the Paria. Each year, these tributaries together contribute an average of 12 million tons of sediment—enough to fill a line of standard 16-ton dump trucks standing bumper to bumper for a distance of over 400 miles. Yet, while this may seem like a big haul, it only accounts for about 15 percent of the former load of sediment through the Grand Canyon.

More important than the total amount of sediment, though, is the quantity of sand donated by these tributary rivers. Finer-grained materials such as silt and clay are so light that they remain suspended in the water column; only sand rebuilds the bars and beaches of the river, and sand is in short supply. Less than 25 percent of the sediment the Paria and Little Colorado Rivers contribute is sand; the rest is mostly mud.

Sand is not the only thing missing from the postdam Colorado. Also gone are the floods necessary to rebuild the beaches and sandbars. The old Colorado was a wild and rowdy river whose flow could vary an astounding 600 percent from one year to the next. Tremendous spring floods fueled by snowmelt in the river's Rocky Mountain headwaters scoured sediment from the channel. Redepositing it above waterline, the floods created beaches, sandbars, and streamside habitat for riparian flora and fauna, as well as areas of slack water inhabited by native fish. The sandbars' highest areas lie above the peak water level released from the dam, so it is impossible to rebuild them under normal dam operating conditions. Once Glen Canyon Dam's gates slammed shut and Lake Powell began to fill, the unruly river's flow became regulated, and seasonal floods no longer occurred. In the absence of these restorative floods, the high areas have either been eroded by low flows and the wind or have been quickly overrun with invasive vegetation. Among other effects, the erosion has exposed and destroyed archaeological artifacts, reduced camping spots for the multimillion-dollar rafting industry, and cut endangered fish species off from critical spawning grounds.

The Colorado's largest historical flood occurred in 1884. To calculate just how large it was, hydrologists donned detective hats and pieced together a variety of clues involving Lees Ferry, a cat, and an apple orchard. We will follow in their footsteps by visiting the old orchard that in 1884 was tended by ferryman Warren Johnson and his family. From the pullout above the rivers' confluence, drive 0.5 mile east to a fork at the bridge over the Paria. Veer left onto the dirt road, which terminates after 400 yards at the Johnson homestead and orchard. Standing in the orchard with the Colorado River barely visible, it's hard to believe that in 1884 the river actually flooded this area, driving the family cat far up one of the trees. Years later, Johnson still remembered how high his cat had climbed, and using this unconventional data point, geologists estimated that the river's flow reached a whopping 300,000 cubic feet per second. In other words, a volume of water that would cover a football field to a depth of ten feet flowed past this orchard every second. Incredibly, flood deposits downstream from here offer evidence of even larger floods, perhaps as much as 500,000 cubic feet per second, in the last 1,600 years.

To see the types of deposits a large flood leaves behind, backtrack to the bridge over the Paria and turn right, following signs to the small ranger station and visitor center located on top of the low mesa a short but steep 600 yards from the bridge. This mesa is an old gravel bar. Very rounded, fist-sized cobbles deposited by the Colorado River are scattered across the ground everywhere up here. While the predam river could easily raise a flood large enough to leave cobbles such as these high above its normal banks, the postdam river cannot. Glen Canyon Dam has replaced these extreme seasonal fluctuations with smaller daily oscillations corresponding to hourly demands for electricity. Until the early 1990s, the dam produced mostly "peaking" power, valuable on-demand electricity that caused flows to vary up to 30,000 cubic feet per second per day. In some parts of the Grand Canyon, daily "river tides" rose and fell up to 13 feet. When environmental studies showed that these fluctuations caused significant beach erosion, the Bureau of Reclamation, which operates the dam, altered its schedule to minimize fluctuations. Still, the requirements of generating electricity make some daily oscillation inevitable.

Rivers are integrated systems, and changes upstream impact areas far below. With less streamside habitat and no restorative floods, tamarisk, a stalwart invasive species, has heavily colonized those beaches that have not eroded, crowding out native vegetation and reducing biodiversity in an ecosystem that until recently supported 2,000 species. Ironically, the tamarisk has attracted five times more birds than used to live in the Grand Canyon, including at least two endangered, though non-native,

During peak floods, the predam Colorado River was able to transport large debris, like the river cobbles shown here. The biggest of these particles are about 3 to 4 inches across.

species: bald eagles and southwestern willow flycatchers. The presence of these endangered species further complicates the situation for river managers, dam operators, and national park officials as they seek a balance between human needs and desires for water, power, flood control, and recreation, and the needs of the Grand Canyon ecosystem.

Tamarisk grows thickly at Lees Ferry. You can examine some by backtracking to the bridge over the Paria, crossing it, and continuing 600 yards to the boat ramp parking area. Thick clumps of tamarisk are the tallest vegetation at the boat ramp's edge. The only places where it doesn't grow along the water's edge are at the boat ramp itself, where humans have cut it away, and the delta of the Paria River, where natural floods periodically demolish it. The park service has also initiated a revegetation project in which it replaces tamarisk with river willow saplings and other native vegetation. Tamarisk consumes water voraciously, so its prevalence along the Colorado River corridor causes significant water losses for humans as well as for native plants. At a conference in 2005 during a severe drought, the only thing Upper and Lower Basin water managers could agree on as they haggled over dwindling water resources was the expansion of tamarisk eradication programs in an effort to squeeze more water from the river.

Beaches and sandbars are essential components of the Colorado River's ecosystem, and the lack of sand and replenishing floods in the postdam Grand Canyon is a serious issue. An even more fundamental concern, however, is that changes wrought by Glen Canyon Dam have resulted in a collapse of the base of the Grand Canyon's food chain. The predam Colorado's sediment-choked waters delivered huge amounts of carbon from recycled organic matter—disintegrating logs, leaves, and animal remains—gathered from throughout the river's basin. This organic matter sustained a diverse group of aquatic insects and formed the base of a complex food web. Today, this organic matter settles to the bottom of Lake Powell and stays there, and the waters released by Glen Canyon Dam are clear enough for a type of stringy algae known as *Cladophora* to thrive. Aquatic insects now eat this algae, making it the new base of the aquatic food chain. With so much *Cladophora* available, the insect population is higher than it was before the dam. However, the dramatic drop in food diversity has resulted in less diversity among insects. As the fish feed on the insects, and the birds and small mammals eat the fish, the changes gradually reverberate through the entire food chain. A system once nourished by carbon from recycled organic matter is now based on photosynthesis, and while it is clear that this has profoundly altered the ecology of the Grand Canyon, no one knows yet how pervasive these changes will ultimately be.

Despite dramatic changes at the base of the food chain, most publicity has focused on endangered fish, which are considered bellwethers for the entire ecosystem. As you stand at the Lees Ferry boat ramp, you will likely see a number of fishing boats launched. In the postdam river, over a million non-native cold-water fish, including rainbow, cutthroat, and brown trout, now thrive in the altered environment, and the 15-mile stretch from Lees Ferry to the dam is a world-renowned gold medal trout fishery.

Walk over to the boat ramp and stick your hand in the water. Does it feel too cold to swim in? Prior to the construction of Glen Canyon Dam, the Colorado's water temperature varied widely, from 80 degrees Fahrenheit in summer to nearly freezing in winter. Now, however, there is almost no variation. This is because water of different temperatures becomes layered in the reservoir. To generate power, water is drawn into the dam about a third of the way down the dam's face, which happens to be within the reservoir's lowest, coldest layer. This water is too deep for much sunlight to penetrate and heat it and too dense to mix with the warmer layers above, so it remains frigid (averaging 46 degrees Fahrenheit) year-round. In fact, the river is so cold that rafters who swim in the water occasionally

The Lees Ferry boat launch serves both rafters embarking downstream through the Grand Canyon and fishermen heading upstream to pursue gigantic trout in Glen Canyon.

become hypothermic even in July, when the air temperature is well north of 100 degrees.

The introduced fish that thrive in these cold, clear waters prey on juveniles of the native species. That predation, intense competition for food presented by these invaders, disappearance of habitat, and reduced access to warm side-stream spawning grounds triggered by the loss of sandbars have radically reduced the number of native fish. Three of the canyon's eight native species have disappeared entirely, and another three are now threatened. Two of these, the humpback chub and the razorback sucker, which require steady water levels for both feeding and spawning, are being closely watched. In 2002, biologists found just 1,100 adult humpback chubs in the river—an 85 percent decline since 1982; 30,000 is considered a healthy, sustaining population. The Grand Canyon's little-known Kanab ambersnail is also endangered, and even the desert bighorn sheep has lost valuable habitat. Are these warnings that the entire system is on the verge of a collapse? Will a different but equally functional ecosystem take its place?

Ultimately, scientists, politicians, and the public need to understand and apply the knowledge of Colorado River geomorphology and Grand Canyon ecology that has been gained since the gates closed on Glen Canyon Dam in 1963. The mission of the National Park Service is to protect and preserve the Grand Canyon for future generations, a mandate that Congress strengthened with the 1992 Grand Canyon Protection Act, requiring that the dam be operated in a manner that safeguards resources within the Grand Canyon. Many people feel that the consequences of doing nothing are too severe. Yet what are our options?

Millions of people in the American West now depend upon Glen Canyon Dam for water, and the numbers grow each day. A few groups, including the Sierra Club, have staged protests, calling for the dam to be decommissioned and Lake Powell drained, but there does not appear to be any political desire for this to happen. Congress has specifically instructed the Bureau of Reclamation, the agency that built and operates the dam, not to study this option.

Instead, efforts have focused on a novel idea: using Glen Canyon Dam, which was built in part for flood control, to generate a flood. Based on their understanding of the natural flood cycle on the predam Colorado, geologists hoped that sediment accumulated for years along the river's bed would, in a moderate flood, be scooped out of the channel and redeposited above the river's high-water mark, creating and restoring beaches for the flora, fauna, and river-runners to colonize. Scientists hoped that sustained flooding would also restore backwater breeding pools, water native vegetation higher up on the sandbars, and flush non-native fish, which are not adapted to high flows, from side-stream habitat critical to endangered fish.

The first attempt at using the dam to create a "controlled flood" came in 1996. The Bureau of Reclamation cranked open the valves for eight days to let 45,000 cubic feet per second down the canyon, the largest volume of water to flow through it since the emergency of 1983. This flood, however, lasted too long. Sand initially deposited during the first few days was later removed by the erosive power of the final peak flows. Conversely, the flood was not large enough to flush out the non-native fish or to remove much invasive vegetation. Although the disheartened scientists concluded that the experiment itself had failed to achieve its objective, it did provide them with valuable new information that they could use to design a better flood.

The assumption at the time of the initial experiment was that, under normal dam operating conditions, sand from tributaries like the Paria accumulates in the river channel over a number of years, but since 1996 researchers have learned that this is not the case. The postdam river is so deprived of sediment that it is what geologists call "starved"; within just a few weeks or months it carts downstream any sediment it receives from the tributaries, leaving little in the deepest part of the channel for floods to pick up and redistribute onto sandbars. The results of a smaller, four-day flood attempt in 2004 were initially encouraging, but ongoing studies have shown that the Grand Canyon's sandbars and beaches continue to dwindle.

Future options may include additional controlled floods timed to begin within two weeks of natural flooding on the Paria and/or Little Colorado Rivers. Another possibility is augmentation of sediment—literally piping sand and mud from Lake Powell around the dam and back into the Colorado River. However, drought-reduced lake levels as well as possible heavy-metal contamination of sediment accumulating behind the dam raise legitimate concerns about both options.

Increasing the amount of sediment in the river would reduce its clarity, and this, combined with utilization of warmer water drawn from higher in the reservoir, could help revive some of the native fish populations. However, without the nourishing organic matter that sustains the base of the food chain, all of these changes are likely to be only partially and temporarily effective, as the U.S. Geological Survey concluded in a 2005 report. The only way to restore the Grand Canyon's native ecosystem is to remove Glen Canyon Dam, and most people do not consider this feasible.

According to legends of the Hualapai Indians, one of several tribes living along the river, their people sprang from the clays, willows, and reeds along the Colorado's banks. The river is an allegory for life. But today, with its banks and stream so altered, the Colorado reminds us that it is more than an allegory. The river has life of its own, which may wane unless we do something about it. Whether the dam is decommissioned or stays until its reservoir fills with silt, we must live with the consequences.

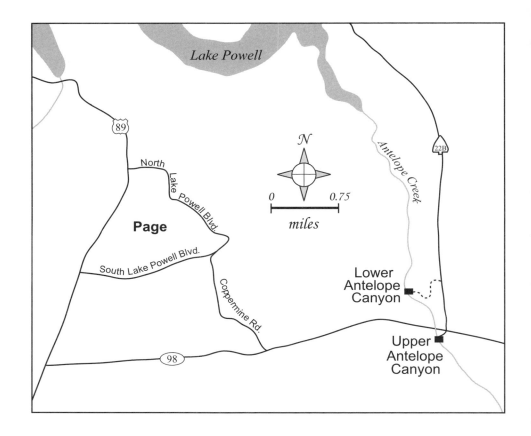

❖ GETTING THERE

Antelope Canyon/Lake Powell Park, a Navajo Nation tribal park, lies just southeast of Page. The park contains two spectacular slot canyons, Upper Antelope (200 yards long) and Lower Antelope (0.25 mile long). To reach the park from U.S. 89 near Page, take Arizona 98 east for 4.5 miles to milepost 299. For Lower Antelope Canyon, turn left 0.5 mile after the milepost onto Navajo Route 22B at the "Antelope Point" sign. Continue for 0.6 mile and turn left onto a dirt road marked by a large homemade sign for the slot canyon. After a few hundred yards, the road ends at a parking area and entrance booth. To reach Upper Antelope Canyon from Arizona 98, turn right 0.5 mile past milepost 299 onto Navajo Route 22B at the sign marked "Antelope Canyon." Follow the road 100 yards to the parking lot. Tribal regulations require that, to enter the upper canyon or get to the lower canyon, you must hire a guide. Several companies in Page offer guided tours, or you can usually find a guide in the canyon parking lot from 8:00 a.m. to 3:30 p.m.

4
Jewel of the Southwest
ANTELOPE CANYON

As soon as you step inside Antelope Canyon, a slot canyon in northernmost Arizona, it envelops you. Sensuously curving walls arch high above you. They form bulbous chambers wider at the bottom than they are overhead, giving the canyon the feel of a cave illuminated by a single sunbeam. Soft light filtering through the narrow roof slit suffuses the complex curves of the walls, bathing their sweeping patterns with an ethereal glow. The interactions between sunlight and the particular orientation and shape of this canyon make for dazzling lighting effects reminiscent of the facets of a diamond, scattering sunbeams and unveiling the brilliant fire within.

After experiencing such a spectacular light show, visitors to Antelope and other slot canyons are overcome with a sense of awe and reverence; many liken the landforms to temples or cathedrals. This stems from the canyons' rarity and their breathtaking intimacy. Stretch out your arms in Antelope Canyon and you can in many places span the gap between the 140-foot-tall walls. An unusual breed of chasm, slots are many times deeper than they are wide; in most canyons those dimensions are reversed. In all its majesty, the Grand Canyon is a full 10 miles wide and "only" 1 mile deep. Even Colorado's spectacular Black Canyon of the Gunnison, one of the world's narrowest gorges, averages just 1.5 times deeper than it is wide, far short of the impressive ratio of Antelope Canyon, which is thirty times deeper than it is wide. Such a constricted chasm is indeed uncommon.

Because access to Lower Antelope Canyon is more technical, with metal stairs and ladders for negotiating drop-offs, we focus on Upper Antelope Canyon in this vignette. True slot canyons are unusual features, even in northern Arizona and southern Utah, which have the highest concentration in the world—a few dozen spread across an area the size of Indiana. The key to comprehending why slot canyons are relatively plentiful in this corner of the globe is to understand how, when, and where they develop.

There are several prerequisites for the formation of a slot. Though the exact recipe can vary, the basic ingredients include resistant rock, water, and a youthful landscape. Not surprisingly, only certain rock types are well suited to forming slot canyons. First and foremost, a nearly perfect layer must be exposed. Like a diamond in the rough, this raw material must be uniformly hard and strong in composition, with few weaknesses or flaws.

Because its grains have been patiently sorted by wind, desert dune sandstone satisfies these stringent requirements. In particular, the Navajo sandstone, seen in the walls of Antelope Canyon, comprises one of the thickest layers of pure sandstone found anywhere on earth and is thus ideally suited to host slot canyons.

Deposited between 192 and 178 million years ago, the Navajo sandstone is the remarkable legacy of a great sand sea that once stretched from here to modern-day Wyoming. Comparable in size to the Sahara, this ancient desert hosted sand dunes that towered a 1,000 feet high. The

Artfully sculpted walls of Upper Antelope Canyon. Note the crossbeds and bounding surfaces in the sunlit back wall.
—Fred Hanselmann, Rocky Mountain Photography (Hanselmann photography.com)

sweeping lines you see in Antelope Canyon's walls are a direct inheritance, now frozen in time, from these ancient dunes.

As you wander up the canyon, take a moment to examine some of the thousands of individual laminations, each thinner than a sheet of cardboard, that comprise Antelope Canyon's walls. Notice how they seem to come in groups, with one package tilting crazily one way, and a separate group just above or below slanting in a different direction. Geologists call each tilted lamination a crossbed and each package of parallel crossbeds a cross-set. Nearly horizontal lines called bounding surfaces separate cross-sets from one another. In this area, the Navajo sandstone blankets the land horizontally, parallel to the bounding surfaces, largely undisturbed by the tectonic forces that have swept this region over the last 192 million years. So if the Navajo has not been tilted by tectonic uplift, what accounts for these anything-but-horizontal crossbeds?

It turns out that, unlike most sedimentary layers, they were deposited at steep angles; each of the thousands of crossbeds you see on Antelope Canyon's walls preserves a single sand avalanche down the inclined face of an ancient sand dune. If you've ever frolicked in a field of dunes, you

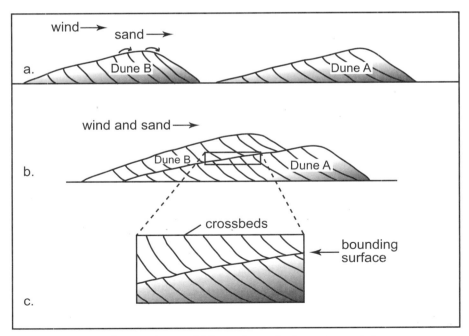

Bounding surfaces, which are formed when one sand dune planes off the top of another, separate set of crossbeds.

know that the downwind sides are steep enough to gallop, jump, and tumble down. In contrast, the upwind side is much gentler. Wind blows sand up this gentle face and piles it at the dune crest, where eventually it cascades down the steep lee face in a turbulent sand avalanche. Through the processes of lithification, crossbeds are locked in stone. Just by looking at their orientations, you can still tell which direction the wind was blowing in this spot over 190 million years ago.

As you admire Antelope Canyon's walls, focus on the cross-sets, the packages of crazily tilting laminations. Sand dunes are always at the wind's mercy, and individual dunes last only as long as the wind keeps blowing in the same direction. Each cross-set represents the migration of a single dune before a major shift in wind reshapes the sand into a new dune and drives it in a different direction. As the young dune begins to travel, it planes the top off the previous dune, leaving a flat boundary, or bounding surface, between cross-sets. You can see many such cycles stacked on top of each other in Antelope Canyon's walls, allowing you to actually count the number of dunes that passed this way during the life of the desert.

Tilted crossbeds truncated by a bounding surface

Wind is one of the most efficient natural mechanisms for sorting sediment. Because wind isn't strong enough to pick up large particles, it only transports sand and dust. Any obstruction the wind meets will slow it down, causing heavier sand to drop to the ground while lighter dust keeps blowing by. In a self-reinforcing cycle, a bush or small bump can force wind to slow and drop its sand, causing the formation of a 1,000-foot-high dune. Because the Navajo sandstone was deposited by many generations of sand dunes, it now comprises one of the thickest layers of nearly pure sandstone found anywhere on earth—over 2,000 feet thick in places.

At the time the Navajo sandstone was deposited, North America's tectonically triggered wanderings placed the continent in the tropics, where the dominant trade winds blow from east to west. The massive bulk of the Pangaean supercontinent, which at that time was just barely beginning to split apart, lay to the east, robbing the trade winds of almost all their moisture before they reached northern Arizona on Pangaea's west coast, leaving it a parched desert. Recent research has shown that periodic climate fluctuations occasionally brought monsoon rains to the area, nourishing a series of small, far-flung oases, which were sanctuaries for a surprising diversity of plants and animals. Fine-grained mudstones and some limestones accumulated in these oases, rare blips in the otherwise monotonous 14-million-year-long accumulation of sandstone. These oasis deposits play a key role later in our story.

In the recipe for slot canyons, once the perfect rock is in place, water is required to actually carve the canyon. This erosion occurs mainly during large floods, when terrifying walls of water roar between the narrow walls, sweeping away everything in their path.

Antelope Canyon has hosted many such deluges. The most recent occurred on August 12, 1997, when a torrent of water 10 to 50 feet high surged through the canyon, tragically sweeping eleven hikers in the lower canyon to their deaths. As you walk through the slot, you will occasionally see logs high above you, wedged between the narrow walls. These are graphic illustrations of the immense volumes of water that flush through this canyon during these brief but violent floods.

Water uses a variety of tools to carve slot canyons. When a river is flooding, its swollen waters lug huge loads of sand and gravel. Most of this material is dragged along the riverbed, literally sandblasting the channel and deepening the canyon in the process. Some of this debris also gets hurled at the walls, eventually widening them. However, because most of the abrasion occurs along the channel bottom, widening is a slower process, and the slot deepens much faster than it widens.

As a flash flood enters Antelope Canyon, the water, which in the wider channel upstream may have flowed a few feet deep, almost instantaneously rises many tens of feet as it is squeezed into the extremely narrow slit. Like wind racing through a tunnel, this venturi effect accelerates the water, creating incredible turbulence in the gorge. Sand and gravel that would normally be dragged along the bottom are boosted high into the water column, where they forcefully pelt the walls in turbulent eddies, scouring out the marvelous scallops that adorn the canyon walls.

    In a cycle that feeds on itself, an eddy at one location enlarges the cavity a fraction. In the next flood, this area becomes the focal point of even more turbulence and continues to widen, eventually excavating the distinctive, bulbous chambers of the canyon. Furthermore, as Antelope Canyon deepens, it holds more water, which increases both the water velocity and the erosive fury it directs at the slot's lower walls.

    Under tranquil conditions, the water barely has enough power to pluck grains of sand from the rock and erode it. During a flood, however, the water becomes filled with frothing air bubbles that implode against the walls at hypersonic speed, unleashing enough force to slice steel. Called cavitation, this force works in concert with sandblasting to carve and polish Antelope Canyon's graceful facets. Eventually, however, the same processes that have fashioned this jewel will lead to its demise. This is why the final basic ingredient for carving a slot canyon is a youthful landscape no more than a few million years old.

    As the landscape evolves over time, the slots, carving deeper, slice through weaker layers either below the Navajo or within it (the oasis deposits). The encounter between the rushing floodwaters of a slot and weak rock layers quickly spells the end of the slot thanks to water's inherently wandering nature. For a variety of reasons, some of which scientists don't understand, water does not like to flow in straight lines; instead, it likes to meander. This is undoubtedly due at least in part to irregularities in the riverbed's topography, but controlled experiments have shown that even in the absence of small deviations, water begins to wander. An urge to be aimless seems built into its very essence.

    Once a meander starts, it inevitably grows, similar to the sand dunes, in a self-reinforcing manner. The water on the outside of the bend flows most swiftly, and swift water has more erosive power. This causes rivers to eat away at the material on the outside of each bend, and as a result, the meander grows over time. The more resistant upper walls begin to overhang at river level, eventually losing support and collapsing, thus widening the gorge. Once this process begins in earnest, the river widens its canyon much more rapidly than it deepens, quickly erasing any slot canyons that may once have existed along its course.

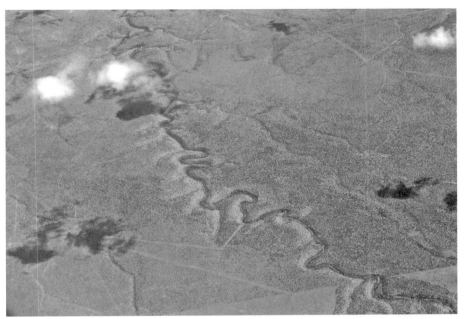

Aerial photograph of a meandering river. Horseshoe-shaped oxbow meanders developed and grew due to the greater erosive power of the faster-moving water on the outside of each meander bend.

As you approach Upper Antelope Canyon from the parking area, its most distinctive feature is the abruptness with which it ends (since you enter at its downstream side). All slot canyons terminate where the stream has breached softer rock that is more susceptible to the power of the abrasive-filled floodwaters. In this region, this usually means that the stream has either cut clear through the thick Navajo sandstone into the underlying, softer Kayenta formation, or it has encountered a weaker layer within the Navajo itself, usually consisting of the finer-grained oasis sediments—soft chinks in the Navajo's otherwise homogeneous sandstone armor.

Unfortunately, modern flash-flood deposits dropped at Antelope Canyon's exit obscure the bedrock channel, preventing us from directly observing which rock type is present. However, the presence of Lower Antelope slot canyon just a short distance downstream suggests that the upper slot ends not in the extensive Kayenta formation, but rather in a weak interval in the Navajo sandstone itself. Walking from Lower to Upper Antelope Canyon, you may be treading on an ancient desert oasis.

The dramatic approach to Upper Antelope Canyon, at its downstream end, where the slot ends abruptly

If these oasis sediments had not accumulated, Upper and Lower Antelope Canyon would likely be connected in one very long slot.

Upper Antelope Canyon begins as abruptly as it ends. After hiking through to its upstream side, take a moment to scramble up the banks of the wash and gaze down on the slot. From this perspective, it is obvious that the canyon has been carved along a straight line. The same is also true for Lower Antelope Canyon.

Like almost all of the Colorado Plateau's slots, both of these canyons are linear because they were carved along a joint. Joints are long, planar hairline cracks that develop in the rock in response to tectonic stresses, contraction during cooling (for igneous rocks), or expansion, as erosion relieves deeply buried rock layers of their immense burden of overlying material.

At the upstream end of Upper Antelope Canyon you can see the linear nature of the joint along which the canyon has been cut. The slot is just a narrow, straight slit at the canyon rim, but becomes wider and more bulbous below due to scouring by flash floods.

All rocks have multiple sets of parallel joints in them, and the twin forces of weathering and erosion exploit these preexisting weaknesses. First, freeze-thaw cycles and other small-scale weathering processes widen the cracks, helping to concentrate water flow. Once water begins flowing down the joint, it accelerates the erosion, eventually turning the crack into a full-fledged stream course. Because linear joints play an important role in focusing stream flow, young drainage networks often have a gridlike appearance in which streams follow one joint for a stretch before shifting to a perpendicular one, then on to the next in a boxy pattern known as a rectangular drainage. As a landscape matures, water's meandering nature frees itself of the joints' linear confines, developing into a more familiar dendritic network of streams resembling the branching veins of a leaf.

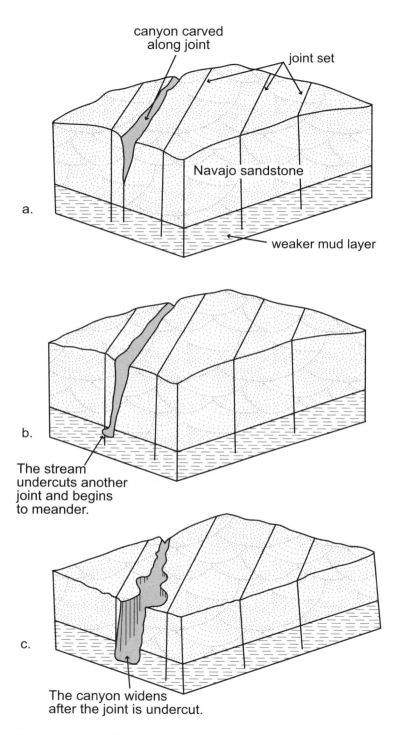

Slot canyons often develop along a joint in a relatively strong rock type but then are destroyed by undercutting of adjacent joints, which causes rockfalls that widen the canyon.

rectangular drainage pattern	dendritic drainage pattern
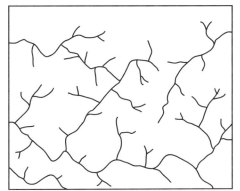	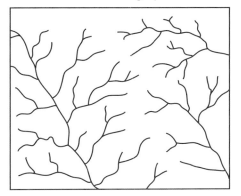

Rectangular drainage patterns usually follow sets of joints in young landscapes, whereas the dendritic drainage networks characteristic of more mature landscapes largely ignore the pattern of joints.

Joints favor slot canyon development because they act like guide grooves for a crosscut saw, focusing the flow of water and its abrasive load of sand and gravel. Joints concentrate the erosion as long as the host rock is strong, thick, and uniform enough to contain the river.

The Colorado River is the master stream that sets the tempo for all other river erosion on the Colorado Plateau. Recent research indicates that Glen Canyon, the 800-foot-deep canyon on the Colorado River just north of Antelope Canyon, which is now flooded by the waters of Lake Powell (see vignette 3), was carved in just the last 500,000 years. Because Glen Canyon established a new low-elevation point in this area, it triggered a host of new tributary streams that funnel water from the surrounding Navajo sandstone tableland down to the bottom of this canyon—in all, a youthful landscape indeed.

For many of these new tributaries, including Antelope Creek, joints provided the fastest and most expedient paths down to the river. Even in this new landscape, though, weaknesses in the confining walls have allowed most of the drainages to meander free of their linear confinement. A few streams, however, retain their courses along the original joints. A handful of these have carved awe-inspiring slot canyons in the process, including Antelope Canyon. Unlike diamonds, which jewelers boast will last forever, these jewels of the Southwest are ephemeral, rare and beautiful to behold.

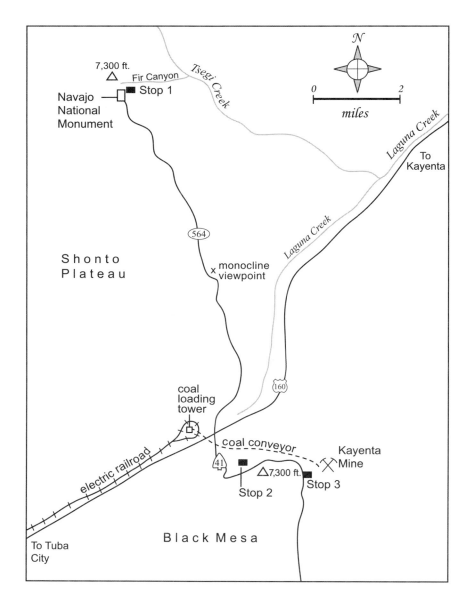

❖ GETTING THERE

From Kayenta, travel west on U.S. 160 to its junction with Arizona 564 and Navajo Route 41 (0.7 mile past milepost 375). Turn right (north) onto Arizona 564 for 9.3 miles to the Navajo National Monument visitor center. Follow the 0.5-mile Sandal Trail behind the visitor center to stop 1, a spectacular overlook of Betatakin ruin. To get to stop 2, an overlook on the north rim of Black Mesa, return to U.S. 160, cross the highway onto Navajo Route 41, and continue south 1.8 miles to a large gravel parking lot on the left, at the crest of the mesa. In this vignette, we orient you to some sights you pass between the first two stops. Stop 3 is at the entrance to the Kayenta coal mine, 4.5 miles farther along Navajo Route 41 (0.3 mile past milepost 6). Turn left into the mine entrance road and park in the pullout on the right just before the gate.

5

A Tale of Twin Plateaus:
NAVAJO NATIONAL MONUMENT AND BLACK MESA

Twenty miles west of Kayenta, an imposing highland stands athwart U.S. 160's diagonal transit across northeastern Arizona. A lone canyon carved by Laguna Creek cleaves the flat-topped plateau in half, offering the highway easy passage through the formidable barrier. The northern half of the highland is known as the Shonto Plateau, and the southern segment is called Black Mesa. Both plateaus top out at just over 8,000 feet in elevation, rising over 2,000 feet above the surrounding countryside. Both climb through several ecosystems, from scrubby grasslands to their pine-clad crests. Despite these similarities, though, the underlying geology of these two plateaus is entirely different. They shared a history through most of geologic time, but about 70 million years ago their paths began to diverge. Though their stories have deviated, the plateaus remain inextricably linked. This linkage continues today as natural resources provided by the plateaus fuel the economies of the Navajo and Hopi tribes. But extracting the resources has provoked difficult questions regarding economic development, environmental protection, and cultural heritage for these longtime inhabitants.

We will begin this tale from the top—the top of the Shonto Plateau; specifically, at Navajo National Monument, a place of cultural significance and profound beauty. The monument's centerpiece is the spectacular Betatakin cliff dwelling, of which you get a superb view from stop 1. The setting for this ancient city is simply breathtaking—an unusually large alcove in the steep, salmon-colored cliffs of Fir Canyon, a small tributary of Tsegi Canyon. Tsegi Creek is the main creek that drains the plateau. Betatakin, home to Ancestral Pueblo people (formerly called Anasazi), consists of several tiers of buildings scattered across the alcove. Tree-ring dating of timbers from these structures indicates that it was built about A.D. 1250, fifty years before the Ancestral Puebloans abandoned the region for reasons archaeologists still debate.

Fir Canyon's stunning cliffs are carved almost entirely from the Navajo sandstone, a thick deposit of well-sorted sand laid down in a giant dune field 192 to 178 million years ago, during Jurassic time (see vignette 4). The Navajo sandstone, which you walked across on the trail from the visitor center to the overlook, forms the bedrock surface of the Shonto Plateau. Look trailside for its sweeping, large-scale crossbeds; these

The Ancestral Puebloan city of Betatakin spreads across the base of a gigantic alcove created by a line of springs that eroded the Navajo sandstone.

prominent features are the hallmark of deposition in desert dunes. If you look below and to the right of the ruin, you can glimpse lighter, thinner horizontal beds of sandstone and mudstone near the floor of Fir Canyon. This is the top of the Kayenta formation, which was deposited by a series of streams flowing across a broad floodplain during a comparatively wetter period earlier in Jurassic time.

The Navajo and Kayenta formations are part of a thick and diverse stack of sedimentary layers that accumulated across the region throughout most of Paleozoic and Mesozoic time. For hundreds of millions of years, this area lay at or near sea level, and sediments accumulated in settings fluctuating between offshore, coastal, and onshore environments, depending on the exact height of the sea. One ocean after another flooded the region, depositing marine mud and limestone. When each sea retreated, the area was exposed as a coastal plain crisscrossed by lethargic rivers or swept by coastal dunes. Gradually, the stack of sedimentary layers built to a prodigious thickness of about

8,000 feet, burying older Precambrian granite and metamorphic rocks far beneath the surface.

About 70 million years ago, compressional forces began to squeeze the entire Four Corners region in a great mountain-building episode known as the Laramide Orogeny. This event severely crumpled the earth's crust in Colorado and New Mexico to form the Rocky Mountains. The Colorado Plateau was also created, but in a different way. The orogeny simply lifted the plateau thousands of feet high. While we often see sedimentary layers tilted on their sides or folded by uplift, this lifting was strangely passive: the vast majority of the layers remained almost as flat as the day they were deposited.

There were, however, a few significant exceptions to this elevator-like rise. The region's Precambrian basement rocks had endured over a billion years of tectonic insults, including much folding and many faults, before the first Paleozoic sediment was ever laid down. By Laramide time, these tectonic upheavals had long subsided, and the faults had gone quiet. But an old fault is like a cut on your knuckle; although it may begin to heal, any disturbance can rip it open again. The Laramide Orogeny was just the disturbance that several dozen dormant faults beneath the Colorado Plateau needed in order to rumble back to life. One of these faults lay beneath the Black Mesa–Shonto Plateau area, directly beneath the spot where Laguna Creek would later cleave the broader plateau in half.

Rudely awoken, this steeply tilted fracture began to move, sliding its northwestern side upward relative to its other half. Faults can slip—that is, the two sides slide past each other—at different speeds. Fast fault movement generally fractures overlying rock layers like dry twigs. When fault movement is sufficiently slow, however, the overlying layers can instead bend, like a green sapling you try to break across your knee. Laramide uplift along the fault beneath Laguna Creek was of the slow variety, so most of the sedimentary layers above it flexed like saplings, creating a large, single-sided fold known as the Organ Rock monocline. We'll glimpse this monocline ahead. Ultimately, by 40 million years ago, the rocks of the Shonto Plateau, which occupied the fault's rising side, had been lifted several thousand feet above their counterparts on Black Mesa.

When movement on the Laguna Creek fault ground to a halt, the Shonto Plateau stood as a plateau on a plateau, a local highland that rose substantially above its surroundings on the broader Colorado Plateau. Weather is always harsher in the highlands, so erosion unceremoniously stripped away the rocks at the top of the sedimentary stack, reducing the Shonto's height and exposing the Navajo sandstone you see today.

During a monocline's rise, a narrow section of its rock layers is steeply tilted, while the rocks to either side remain flat, making each

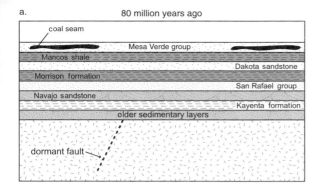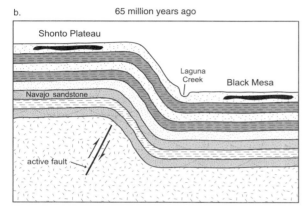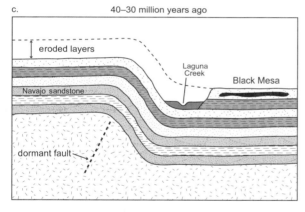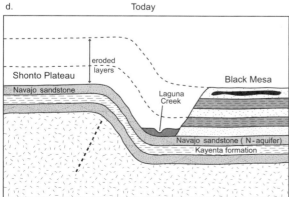

The development of the twin plateaus over time: a) deposition of the stack of sedimentary layers; b) uplift of the monocline leaves the Shonto Plateau higher than Black Mesa; c) Laguna Creek carves its canyon and erosion lowers the Shonto Plateau; d) continued erosion produces today's twin plateaus.

layer resemble a swath of carpet draped over a stair riser. Here on the Shonto Plateau, the Navajo sandstone lies nearly flat along the top tread before abruptly diving into the ground on the plateau's southern side. The layers then gradually flatten out to form the bottom tread, which runs beneath Black Mesa.

Laguna Creek formed at the base of the stair riser, and before long, it had carved down through the rocks far enough to encounter a soft layer. Once in this easily eroded layer, the creek's path of least resistance was to follow it to the east, where it ran along the steeply tilted stair riser between the top tread of the Shonto Plateau to the north and the bottom tread (which would later form Black Mesa) to the south. As Laguna Creek excavated this layer, it entrenched itself along the valley now occupied by U.S. 160. As the creek continued to erode deeper into the monocline, down-dropped Black Mesa, which originally formed the area's lowest point, was gradually transformed into today's towering mesa. Once Black Mesa became a high point, erosion also began to strip its upper layers off, but the substantial head start on the Shonto Plateau ensured that the rocks exposed there would always be older than those capping Black Mesa.

The upshot of this geologic history is that at Betatakin overlook, at an elevation of 7,300 feet, you are standing on the Navajo sandstone, but when you drive across the valley and up to the same elevation on Black Mesa at stop 2, you will be standing on much younger rocks. There, the Navajo sandstone lies thousands of feet below you—where, by the way, it serves a vital purpose for Black Mesa's residents: it's the principal source of water for that parched plateau.

The area's geology dictates that rain falling on the extensive Shonto Plateau necessarily lands on the Navajo sandstone. With a high degree of permeability, or interconnected space between individual rock grains (even if it looks like completely solid rock), the sandstone soaks up the moisture like a sponge, and gravity pulls the water down through its well-connected network of pores. When the water reaches the underlying Kayenta formation, however, it encounters a thick layer of mudstone. Mudstones consist of clay minerals shaped like elongated plates. Because of their shape, they stacked up like shingles as they settled on the Kayenta floodplain. As a result, the Kayenta mudstones are unable to transmit groundwater at anywhere near the same rate at which it moves through the permeable Navajo sandstone above. Water therefore pools up in the Navajo, making the formation an aquifer—a water storage unit.

With nowhere else to flow, the water migrates along the top of the Kayenta in whichever direction is downhill. Here at Betatakin, "downhill" is the floor of Fir Creek. Water stored in the Navajo began to flow out of

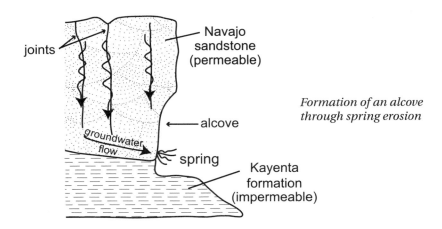

Formation of an alcove through spring erosion

the ground along the base of the canyon wall, forming a line of springs. One by one, the force of the exiting water plucked sand grains from the wall, forming a small recess that gradually enlarged into the one you see today—one of the biggest alcoves on the Colorado Plateau and the perfect location for constructing an Ancestral Puebloan city.

Although some of the water in the Navajo sandstone aquifer comes to the surface in the creeks and washes of the Shonto Plateau, the majority remains underground and continues to flow downhill along the Navajo-Kayenta boundary. Because of the tilt of the Navajo sandstone and its companion layers along the monocline, downhill is toward Black Mesa. There the water collects in a large aquifer saturating the Navajo sandstone. It is this water that numerous wells have tapped for human needs.

As you head down the Shonto Plateau back to U.S. 160, notice that you remain on Navajo sandstone throughout your drop. This is because you are driving down a monocline. If you look east (left) from just about anywhere along this descent, you will be able to trace the Navajo and Kayenta layers as they change from horizontal on the crest of the plateau to steeply tilted approaching Laguna Creek. An especially good place to see this is from the parking area 0.3 mile past milepost 378 on Arizona 564, looking to the left of the road.

When you reach U.S. 160, cross it onto Navajo Route 41, which leads up Black Mesa. A tall green tower to your right (west) is the loading station for trains carrying coal mined from the top of Black Mesa to the Navajo Generating Station at Page, Arizona. The coal will figure prominently in

Layers of Navajo sandstone, the light-colored rock in the distance (left), arch down along the monocline off Shonto Plateau. Shonto's flat top appears far left, and Black Mesa forms the highland on the right. Laguna Creek valley separates the plateaus.

our story at the third stop. Meanwhile, in this valley, the Navajo sandstone disappears beneath the ground, plunging below Black Mesa. As you climb onto the mesa, all of the layers you see are younger than the Navajo formation's 178 million years.

The first layer you will be able to identify is the Morrison formation, the youngest of northern Arizona's Jurassic-aged rocks. Because it consists primarily of soft mudstone, the Morrison weathers easily, breaking down into mud that forms soil, obscuring the rock, and supporting vegetation. Where exposed, the Morrison forms a heavily gullied slope in a variety of pastel hues. The best place to see it is in the roadcut on the left 0.5 mile from U.S. 160, where the formation underlies the row of sandstone cliffs on which the coal conveyor belt crosses the road. The Morrison mud was deposited on vast floodplains created by rivers meandering from a young mountain range known as the Sevier Highlands. The highlands had been recently uplifted to the west in present-day Nevada (more on that in vignette 10). The Morrison is a widespread formation;

its name derives from exposures near Morrison, Colorado, a suburb of Denver. Historically, the Morrison has been an important source of uranium ore throughout the Four Corners region.

The band of sandstone cliffs above the Morrison and below the coal conveyor is the Dakota sandstone. With the Dakota, we move from Jurassic into Cretaceous time, beginning 146 million years ago. This band of well-cemented, resistant sandstone tilts noticeably down to the south along the monocline, but here on the north flank of Black Mesa we are nearing the flat lower stair tread, so the tilt is not as steep as that of the Navajo sandstone you saw as you came off the Shonto Plateau. The Dakota is an even more widespread rock unit than the Morrison; its name stems from outcrops in Dakota County, Nebraska, hundreds of miles northeast. It was deposited as beach sands and in lagoons on the shore of yet another of the area's encroaching seas. This sea was different, however. Whereas all earlier seas had crept in from the west, the Dakota sea entered Arizona from the east, marking a profound shift in the paleogeography of the American West. During the time the Dakota was deposited, the recently uplifted Sevier Highlands shielded this area from the sea to the west. But with global sea level at its highest in at least 500 million years, a shallow sea flooded the midsection of North America, linking Hudson Bay to the Gulf of Mexico. The Dakota sandstone you see here was laid down on the western shore of that vast interior sea.

Above the Dakota cliff, a broad, thickly vegetated trough has formed in the next layer, the Mancos shale. Looking east from the Dakota cliffs, you can see this shale in the distance, where it forms a barren gray slope. Named for exposures in Mancos, Colorado, near Durango, the Mancos is a marine mudstone deposited during the Cretaceous sea's high-water mark. Its drab gray color derives in part from its high organic carbon content, a legacy of the life that teemed in its warm waters.

At stop 2, look across the road to see an exposure of Black Mesa's caprock, the Mesa Verde group. This group consists of a bundle of three related rock formations all deposited in beach, river, lagoon, and coastal-swamp settings when sea level in the interior ocean began to fall. This exposure is typical of the Mesa Verde group, with the blond beach and river sandstones forming resistant cliffs separated by thin, crumbly beds of coal and black shale from the times when floodplains, swamps, and lagoons covered the area. The most prominent black-shale layer lies across the road, on the downhill side of the parking area.

Turn your attention now to the panorama that unfolds to the north. The salmon-colored Navajo sandstone blankets the top of the Shonto Plateau at the same 7,300-foot elevation at which you stand at stop 2. The steep down-flexing of the Navajo and underlying layers along the heart of

The coal conveyor descending from Black Mesa as seen from stop 2. The hummocky, light-gray ground below the conveyor is Mancos shale. Beyond that, a prominent tilted cliff of Dakota sandstone juts out over Laguna Creek valley. In the background, the Navajo sandstone arches up along the Organ Rock monocline onto the Shonto Plateau.

the monocline is apparent north of Laguna Creek. In the middle ground you can easily spot the more gently tilted Dakota sandstone, whose moderate angle is due to its position near the north edge of the lower (Black Mesa) stair tread, where the layers begin their transition back to horizontal. Behind you, in the roadcut, the Mesa Verde group rocks are essentially horizontal, indicating that you are now standing on the flat lower stair tread south of the monocline.

The coal conveyor drops down off the mesa a short distance to the east, and its hum fills the air day and night. The coal it carries comes from a huge strip mine, the Kayenta Mine, which we will see at stop 3. Peabody Coal Company holds mining leases at this and another mine covering a total of 14,000 acres of Navajo and Hopi land. The mines lie in a politically sensitive area claimed by both groups, who have been uneasy neighbors for centuries. When the U.S. government placed the tribes on reservations, it reluctantly assumed the responsibility of adjudicating this centuries-old dispute. Even today, the issue is not completely resolved, but a 1962 court ruling designated the middle portion of the mesa as a joint-use area for both tribes. A 1977 ruling revised that arrangement, but left mineral royalties to be split jointly. These court rulings led to the forced relocation of many Navajo and fewer Hopi, a bitter and protracted episode that continues to the present. High economic stakes further heighten the social tensions.

At stop 3 you can see how a strip mine works. As the name implies, strip mining entails the complete denudation of the land to extract a horizontal seam of coal, in this case about 3 feet thick, a few tens of feet below the surface. First, workers remove the topsoil, setting it aside for replacement during the site's eventual reclamation. Gigantic machines called draglines then remove the overlying rock layers, exposing and mining the coal. Draglines are the most massive land vehicles ever designed by humans. At least one should be visible from this vantage point, although here they are usually far enough away that it is hard to appreciate their true enormity. Not counting their huge boom arms, the largest draglines stand a staggering nine stories tall and weigh in at 4,000 tons. Their buckets can scoop up 100 tons of coal at a single bite. When this machine has extracted all the coal in a place, it ponderously creeps forward to a new area. Other workers and machines come in behind to fill the hole, replace the topsoil, and seed the land in an attempt at reclamation. Native vegetation here consists of piñon and juniper trees. You can see it in the undisturbed landscape to the west. Such piñon-juniper woodland grows back extremely slowly, so the mine's reclamation efforts focus not on restoration of this ecosystem, but rather on planting grasslands for cattle grazing. You can see some of this reclaimed land to your left (east), between you and the active mine.

The Kayenta Mine harvests coal from ten separate seams in the middle unit of the Mesa Verde group. Coal forms from the remains of land plants—in this case, the organic matter preserved in coastal swamps lining the shore of the Cretaceous interior sea. Swamps are excellent places to preserve organic carbon because the volume of vegetable matter is so huge. As soon as a plant dies, bacteria begin to decompose it, recycling the organic carbon back into the biosphere. The bacterial populations of swamps explode because of the abundant food supply, but population size is nonetheless limited by the availability of oxygen, which the bacteria need to live. The amount of dead vegetation in a swamp exceeds the amount that bacteria can consume, leaving substantial amounts of undecomposed organic carbon to be buried by sediments, where it eventually turns to coal.

The mining on Black Mesa leaves a mixed legacy. Archaeological evidence shows that the Ancestral Puebloans who inhabited Betatakin used

A cranelike dragline sits in the middle of the Kayenta strip mine. Light-colored grasslands in front and behind mark reclaimed areas. At right, a coal loading station stands at the head of the conveyor. The conveyor itself cuts like a road through native piñon-juniper woodland in the foreground.

Black Mesa coal for warmth and to fire pottery. Locals continued to take advantage of this resource on a small scale until 1968, when Peabody Coal Company began commercial operations. Since then, the mines have been tremendously important economic engines for both tribes. Besides employing hundreds of local Hopi and Navajo men and women, they pull in an estimated $100 million a year. Both tribes receive millions of dollars in royalties from coal leases and for the use of groundwater in mining operations. Although the Navajo receive the largest payments, the smaller Hopi tribe is more dependent on its share of royalties, which amount to 33 percent of its annual tribal budget.

These economic advantages have a price. Besides the intertribal tensions that have swirled around the operations, it is well-nigh impossible, despite reclamation efforts, to recreate native ecosystems on the disturbed land. The environmental disruption here is permanent. Of even deeper concern to the Navajo and Hopi people, the land they hold sacred has been altered beyond all recognition. They can no longer consider it the same land that held the bones and spirits of their ancestors. The disruption of tribal spiritual practice caused by this alteration has strained the social fabric of some Navajo and Hopi communities. Many tribal members feel profound ambivalence toward the mines.

Finally, there is the issue of water. Coal from the Kayenta Mine fuels the Navajo Generating Station, the massive electricity-generating plant northwest of here near Page. An electric train fed by the conveyor belt you passed on your way up the mesa transports the coal the 80 miles to Page. A second mine, the nearby Black Mesa Mine, which also operated for decades, closed in December 2005, due basically to concerns about water. Coal from Black Mesa Mine was transported 273 miles to the Mohave Generating Station near Laughlin, Nevada, via an 18-inch-diameter slurry pipeline. Moving the coal this vast distance by slurry consumed 2,000 gallons of water a minute—over 1 billion gallons a year. That water came from wells in the Navajo sandstone deep beneath Black Mesa's surface, or the "N-aquifer," as locals call it.

The N-aquifer is also the area's main source of water for household consumption. Any groundwater system achieves a natural balance between the discharge of water out of the ground at springs and recharge through precipitation falling on highly porous rocks. Almost all of the water in the N-aquifer originally fell as rainwater on the Shonto Plateau, the aquifer's major recharge area. But water moves slowly through the Navajo sandstone. In fact, studies have shown that water takes 25,000 to 30,000 years to travel from Navajo National Monument to the wells at Black Mesa. The present rate of groundwater withdrawal far exceeds the recharge rate here. As a result, the groundwater level beneath Black Mesa has dropped over 100 feet.

Hydrologists worry that, at the current rate of pumping, the aquifer—quite small as aquifers go—could be depleted in the next decade. If that occurs, the Hopi and many Navajo will face a critical water shortage. Already, seven of the fourteen major springs on Hopi lands have dried up, and the flow rate at five more has diminished. Marshes and wetlands on the mesa are now dry, as are several wells and the previously perennially wet Moenkopi Wash. Drought is a regular visitor in this parched land; a recent severe drought may have played a significant role in the drying of springs on Black Mesa. But hydrologists agree that extensive pumping of groundwater here adds significant stress on the fragile water resources.

There is a growing sense of alarm among tribal leaders, who find themselves in the difficult position of trying to preserve jobs and royalty revenues associated with coal mining while simultaneously attempting to preserve the water on which the mesa's population depends. The recent closure of the Black Mesa Mine and the Mohave Generating Station it supplied threw hundreds of Native Americans out of work. The silver lining of the closures would seem to be the easing of groundwater withdrawal, but that is not the case. Peabody Coal has actually proposed to increase its withdrawal of water from the aquifer to wash coal extracted from the Kayenta Mine, a process that reduces sulfur emissions when coal is burned at the power plant. Removal of sulfur would obviously reduce air pollution in the region, but at the high environmental cost of threatening the N-aquifer.

The distinct but intertwined geologic histories of Black Mesa and the Shonto Plateau have endowed the twin plateaus with many precious natural resources. The people who inhabit the plateaus must now face the challenging task of deciding how best to use and preserve those resources for present and future generations.

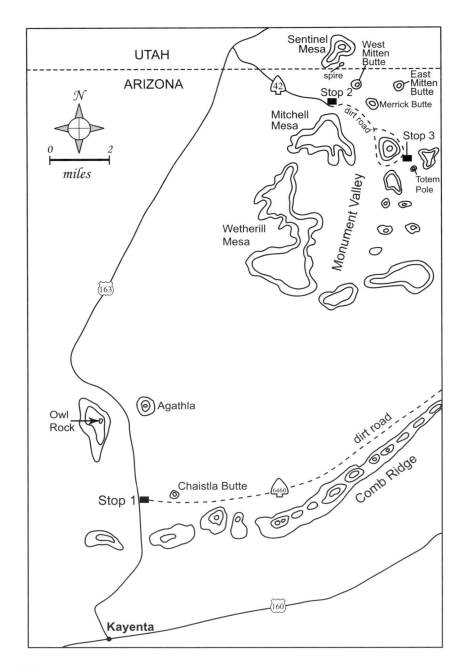

❖ GETTING THERE

From Kayenta, in the northeast corner of Arizona, take U.S. 163 north 5.4 miles. At Navajo Route 6460, an unmarked junction 0.7 mile past milepost 398, turn right (east). Near the junction is stop 1, a pullout on the right where you can soak up the view. En route to stop 1, several drive-by features are noted in the text. To reach stop 2, the Monument Valley Tribal Park visitor center, continue north on U.S. 163 for 23 miles from Kayenta to the Utah state line. Continue 0.4 mile into Utah and turn right (east) at the junction with Navajo Route 42, the Monument Valley road. Continue southeast for 3.8 miles to the visitor center. The

How to Carve a Totem Pole
MONUMENT VALLEY

When legendary movie director John Ford used Monument Valley as the backdrop for his 1938 western *Stagecoach*, this Navajo Nation tribal park in northeastern Arizona became the icon of the landscape of the American West. The valley's steep-walled buttes and slender spires are utterly foreign to the everyday experience of most people on earth.

In this valley of icons, three formations are the most emblematic. The East and West Mitten Buttes and the Totem Pole are the most widely recognized monoliths in the park. The story of their formation is representative of the processes that have formed all of the towers and buttes in this incomparable landscape.

Like most of the scenery on the Colorado Plateau, Monument Valley's beautiful formations owe their origin to a thick stack of colorful layers of sedimentary rock. The sediments from which Monument Valley has been carved were laid down beginning 284 million years ago, when the area was a vast coastal plain crisscrossed by sluggish, meandering rivers whose sources lay in a mountain range far to the east. Upon burial, the mud deposited on the floodplains of these rivers became the Organ Rock shale, the dark brown mudstone that forms the pedestals for Monument Valley's majestic towers. Their sheer cliffs are carved out of the DeChelly sandstone, a pure quartz sandstone preserving the remains of huge sand dunes that blew across the northern Arizona landscape about 275 million years ago. About 27 million years later, the lazy river floodplains were back, depositing more of the red mudstone and sandstone layers of the Moenkopi formation, the first of the Triassic-age layers. After a pause in deposition, more vigorous rivers swept the area for a brief time around 234 million years ago, spreading a thin, resistant sheet of cobbles sheathed in a sandy matrix; known as the Shinarump conglomerate, this is the basal layer of the Chinle formation. These lively, youthful rivers

---

overlook is on the north side of the building. For stop 3, drive 5.7 miles down the 12-mile dirt loop road that winds through Monument Valley. Your destination is the Totem Pole and Yei Bi Chei vista, scenic viewpoint #7 on the map you receive at the park entrance. The road is only open during the day. Allow at least 1.5 hours to enjoy it.

The Mitten Buttes

The Totem Pole—the tallest tower shown here—and the Yei Bi Chei, the group of towers to the left

The rock layers that form the scenery in and around Monument Valley

| Navajo sandstone |
| Kayenta formation |
| Wingate sandstone |
| Chinle formation |
| Shinarump conglomerate |
| Moenkopi formation |
| DeChelly sandstone |
| Organ Rock shale |

soon settled into a sedentary middle age, returning to their sluggish ways a few million years later as they deposited the thick, variegated mudstones of the rest of the Chinle formation.

By earliest Jurassic time, just under 200 million years ago, the climate had dried out again, and sand dunes once more swept across the region's vast plains, leaving the brick-red cliffs of the Wingate sandstone in their wake. Another climate swing then brought more moisture to the region, once more sending rivers weaving across these flats to lay down the Kayenta formation. That moistening proved temporary, though, and by 192 million years ago, drought set in with a vengeance, forming the largest desert ever on the North American continent. This desert was the size of the modern Sahara, and towering sand dunes up to 1,000 feet high covered vast tracts of it. The thick, crossbedded Navajo sandstone is its legacy. Although the deposition of many more sedimentary layers would follow throughout the remainder of Mesozoic time, burying these sediments deep enough to turn them into rock, erosion has ensured that none of those later layers remain today.

The raw materials from which nature would sculpt Monument Valley and its environs were now completely assembled. It was not until 70 million years ago, however, that tectonic and erosional processes began to fashion the modern Monument Valley. At that time, a broad highland area known as the Monument upwarp was uplifted. The upwarp was one of several large domes that rose like welts across the Colorado Plateau during a mountain-building episode known as the Laramide Orogeny, profoundly altering the geography of the American Southwest between 70 and 40 million years ago. The entire Four Corners region was squeezed, thrusting up the Rocky Mountains of Colorado and New Mexico in the process. The Colorado Plateau was also raised thousands of feet, but with little turmoil, that is, hardly any tilting or faulting of rock layers. The exceptions to this elevator-like lifting were found along the flanks of the domes, where slumbering, deeply buried faults awoke in response to the compression and tilted the rock layers into monoclines, one-sided folds resembling a stair step (discussed in detail in vignette 5).

We will now traverse the progression of these colorful sedimentary layers to trace the Monument upwarp from its edges to its core—which was breached to form Monument Valley. Although erosion has produced a slightly downhill trip today, the scenic drive from Kayenta to the tribal park actually takes you up onto the upwarp's crest. As you head out of Kayenta, you pass onto the dome's southern flank just after you cross a bridge over Laguna Creek. At milepost 397, slabs of blond, crossbedded sandstone angling down to the south line both sides of the road, which climbs up a short rise. These slabs consist of 192- to 178-million-year-old

Monument Valley ❖ 83

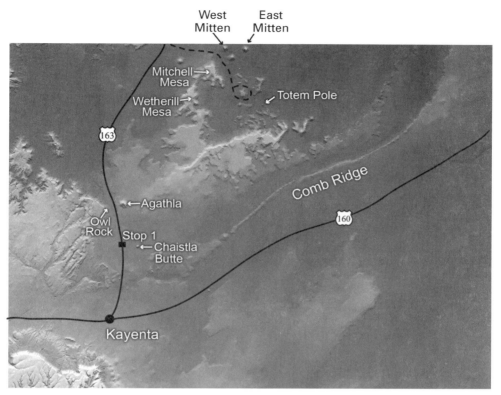

Comb Ridge and U.S. 163 form the edges of the Monument upwarp. Erosion has eaten into the center of this dome, carving out the magnificent buttes and spires of Monument Valley.

Navajo sandstone that has been tilted along the Comb Ridge monocline. The small rise you just climbed is the very western end of Comb Ridge, a dramatic crest of rock arching over the edge of the stair-step fold. In the next mile, the road passes through a small valley in the red Kayenta formation, followed by a second tilted ridge of the underlying pink Wingate sandstone.

From stop 1, in the large valley north of Comb Ridge, you can trace the broad outline of the ancient crustal blister of the Monument upwarp. To the south and east, the tilted Wingate cliff of Comb Ridge forms the skyline along with a nearer ridge made of a resistant layer in the Chinle formation. Comb Ridge, which defines the upwarp's southern and eastern flanks, traces a northeastward arc from here to near Blanding, Utah.

Layers of Chinle formation (left) *and Wingate sandstone* (prominent cliff band on right) *tilt south along the Comb Ridge monocline.*

These tilted layers were heaved up during the formation of the Monument upwarp. Try to mentally trace the extension of the ridges up into the space over your head to get a better sense of the enormous volume of rock that once stood here, supporting a gigantic plateau, before erosion mercilessly stripped off thousands of feet and reduced these cliffs to the stumps you see today. Because the monocline tilted the layers down toward the south, as you travel north along U.S. 163, you plunge ever deeper into the upwarp's eroded core, crossing older and older layers en route.

Two distinct towers separated by the low pass that the highway runs through dominate the view to the north. The western (left) tower is Owl Rock, distinctive because of its bulging shape. Its gothic-looking top consists of a remnant of Wingate sandstone. Notice how this scrap sits conspicuously higher than the outcrops of Wingate in Comb Ridge to your south; this is a concrete demonstration of the northward rise of the rock layers along the Monument upwarp. Below the Wingate cap, Owl Rock's broad pedestal consists of the older Chinle formation, here displaying the variegated mudstones for which it is famous (vignette 15). These Chinle mudstones are poorly cemented, hence they are more easily eroded than the other layers in the sedimentary stack, and in the spot where you are standing, they have been completely removed to form this broad valley between Owl Rock and Comb Ridge.

The taller tower is Agathla, a brooding, gray volcanic neck rising an impressive 1,200 feet above you. Chaistla Butte, the smaller but equally

U.S. 163 runs along a valley eroded in the soft Chinle formation and passes between the towers of Owl Rock (left) *and Agathla* (right).

somber gray tower east of you, is the neck of another volcano that erupted here sometime between 32 and 25 million years ago. Both are part of an expansive group of volcanoes known as the Navajo volcanic field that burst to life across the Four Corners region at that time. New Mexico's famous Shiprock is the largest of these necks; Agathla holds second place.

En route to Monument Valley, you will be able to identify one more volcanic neck by its drab, gray color, which really stands out in this land of vivid reds. These necks are the only remnants of violent eruptions triggered by frothy magmas rich in dissolved gases that originated deep underground, within the earth's mantle. The high gas content caused the magma to rise unusually rapidly and erupt very explosively. The necks themselves were once the conduits through which the magma passed during each eruption. The overlying volcanoes have long since eroded away, but the conduits' more resistant rock stood high as the softer sedimentary layers encasing them were stripped off one by one. From these necks, we know that the general height of the landscape at the time they erupted was at least as high as the top of Agathla, a minimum of 1,200 feet taller than today. It is likely that the upward migration of hot magma during this period of volcanic activity further raised this portion of the Colorado Plateau, making the Monument upwarp even higher.

The two pulses of uplift—first from Laramide compression, then from intruding magma—thus left the future rocks of Monument Valley standing high above their surroundings, setting the scene for the erosional processes that would expose and chisel them into the buttes and towers that delight us today. To see the raw materials that erosion had to work with, we need to continue north, into the older layers in the heart of the Monument upwarp.

At stop 2, at the valley's edge, an amazing view into Monument Valley awaits you. The appropriately named East and West Mitten Buttes, as well as Merrick Butte south of them, dominate the scene. Each butte is primarily composed of a sheer, smooth wall of red sandstone resting on a pedestal of dark, thinly layered mudstone mixed with sandstone. Each wears a scruffy cap of thinner-bedded rock.

This succession of layers picks up where we left off at stop 1. Recall that the valley where we stood before was carved in the soft badlands of the Chinle formation. Those same Chinle mudstones used to lie immediately above the buttes you are gazing at here, but along with the overlying Navajo, Kayenta, and Wingate layers, they have been stripped wholesale from this area. The light-colored cliffs that cap East Mitten and Merrick Butte are remnants of the lowest layer of the Chinle formation, the Shinarump conglomerate. In contrast to the rest of the Chinle, the Shinarump, which was deposited in a series of river channels (which you can walk through in vignette 9), is a hard, resistant layer critical to the creation of the Monument Valley landscape. As you see here, the Shinarump forms a protective caprock for the park's mesas, buttes, and spires, slowing the erosion of the softer layers below and leading directly to the creation of these spectacular rock formations.

If you look carefully at the top of East Mitten and Merrick Buttes, you will see, sandwiched between the topmost Shinarump cliff and the main sheer cliff, a thin section of softer, darker rock that has eroded into a series of ledges and shelves. This is the Moenkopi formation, a sequence of alternating mudstone and sandstone layers deposited on tidal flats and a river floodplain reminiscent of today's Mississippi River delta. Erosion has recently stripped the protective Shinarump cap from West Mitten Butte, leaving just a thin veneer of Moenkopi on its summit.

The dramatic, 400-foot-tall cliffs that are the highlight of all the monoliths in Monument Valley are carved out of the DeChelly sandstone (pronounced duh-SHAY), the same dune-deposited rock that has been sculpted into the beautiful scenery at Canyon de Chelly National Monument (vignette 9). The rubble-covered pedestals beneath the towers consist of the mudstones and sandstones of the Organ Rock shale, which, like the Moenkopi, were deposited by rivers lazing across a vast floodplain. The floor of Monument Valley has been scooped out of the lowermost portions of the very soft Organ Rock shale.

Now that we are familiar with the raw materials, we can explore how nature chiseled these vivid maroon layers into this quintessentially Western landscape. One key lies in our location near the crest of the Monument upwarp, the place that experienced the most extreme tension as the rock layers were flexed up. You can generate analogous forces in a thick slab of cheddar cheese if you squeeze the ends together to form an arch. A closely spaced set of vertical cracks develops right in the cheddar crest. Similarly, as the Monument upwarp grew, the rock units here were all split by vertical cracks that geologists call joints. Although joints are present in all four of the layers that comprise the towers, they are most apparent and abundant in the DeChelly sandstone.

This is because, like most sandstones composed of pure quartz, the DeChelly is extremely brittle. As the stack of sedimentary layers arched skyward, more malleable mudstone units like the Organ Rock shale were able to respond to the stretching through a combination of jointing and plastic flow. The DeChelly, however, did not have that luxury. It accommodated the stress the only way it could, by fracturing along numerous

West Mitten Butte, like all of the monoliths at Monument Valley, is riddled with numerous vertical joints (cracks). A window has formed along a joint on the right side of the main wall. This will likely be the next place the tower is breached, significantly reducing the butte's size.

cracks. These closely spaced, vertical joints later governed the pattern of erosion when weathering processes at last began to carve into the DeChelly sandstone.

The joints in the overlying layers provided handy conduits for water to flow deep into the sedimentary pile, where it expanded as it repeatedly froze during the cold desert winters, shattering adjacent rock in the process. After a bit of widening through this freeze-thaw cycle, many of the cracks provided ideal courses for the streams that drained the landscape after intense monsoon rainstorms. The raging floodwaters following each downpour scoured and abraded these channels, cutting still deeper into the joints. These newly formed drainages breached the resistant

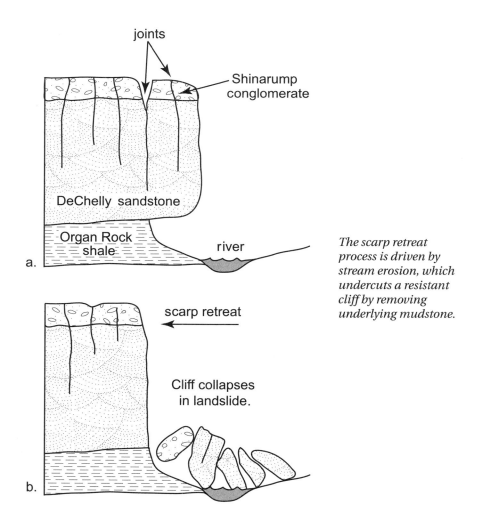

The scarp retreat process is driven by stream erosion, which undercuts a resistant cliff by removing underlying mudstone.

Shinarump conglomerate, quickly sliced through the softer Moenkopi, and became entrenched in the joints splitting the DeChelly sandstone below. Eventually, the streams carved completely through the DeChelly to expose the soft Organ Rock shale. That's when Monument Valley really started to take shape.

When a stream finds itself flowing across a resistant rock layer such as the Shinarump or the DeChelly, most of its power is expended trying to deepen its channel. All streams have a natural tendency to meander back and forth, and along the outside of each bend, where water flows fastest, the sides of the channel are also assaulted by erosion. While this onslaught accomplishes some widening, in strong rocks the rate of downcutting far exceeds the rate of channel widening, forming a deep, narrow gorge (vignette 4). Such was the case here so long as the streams of Monument Valley were slicing through the relatively hard DeChelly sandstone. But as soon as the streams encountered the soft Organ Rock shale, erosion on the outside of each bend began to carve away huge chunks of soft mudstone with each torrential flood, rapidly widening each channel. The towering sandstone cliffs were repeatedly undercut as the Organ Rock shale was literally quarried out from under them. When the space beneath a sandstone overhang cut back far enough to intersect with the next vertical joint, all support for the hanging section of cliff was lost, and it collapsed to the ground in a spectacular rockfall. In this way the cliff bands along all of Monument Valley's washes began to march away from the streams that had destabilized them in a process known as scarp retreat. If you look at the base of each butte, you can see that the shale pedestal is littered with huge slabs of DeChelly sandstone that tumbled off the cliffs above, repeatedly undercut.

Continued scarp retreat gradually shrank Monument Valley's mesas into buttes, and the buttes into spires. The process continues today, and representatives of all three stages are visible from your position by the visitor center. Look north, where the comparatively undissected mass of Sentinel Mesa lies. If you carefully examine its flanks, you will see that a spire has formed adjacent to it as closely spaced canyons have converged, slicing just a sliver off the mesa's southern edge. The Mittens and Merrick Butte to your east are perfect examples of buttes, which form where scarp retreat has sufficiently whittled away at the edges of a mesa to reduce it to a width not much larger than its height—the definition of a butte.

While scarp retreat shrinks mesas into buttes, the freezing and thawing of water in the joints of the resulting monoliths fragments these features from within. The Mitten Buttes provide excellent examples of this process in action. Notice how the thumb of each mitten is a freestanding tower attached to the butte at its base. The gap between the fingers and thumb is slowly enlarging as water percolates deep into the joint at the crook of each thumb. It won't be long—geologically speaking—before more landslides eradicate the thumbs.

The ultimate stage in the inexorable shrinkage of these towers is the formation of a slender spire that is much taller than it is wide. No better

HOW TO CARVE A TOTEM POLE

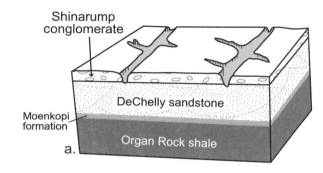

a) Stream courses form along joints in the hard Shinarump conglomerate.

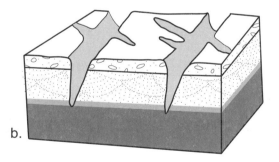

b) Those streams entrench themselves in deep gorges carved through the DeChelly sandstone.

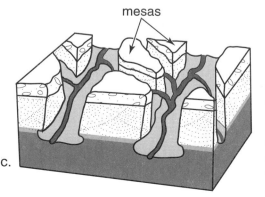

c) As the streams cut down into the softer Organ Rock shale, they widen their floodplains, chopping the overlying layers into separate mesas.

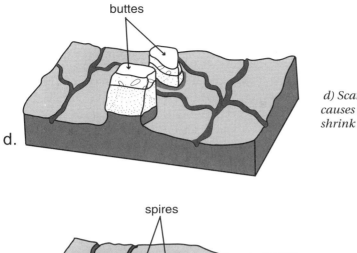

d) Scarp retreat causes the mesas to shrink into buttes.

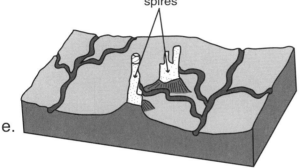

e) Continued erosion carves the buttes into spectacularly slender spires such as the Totem Pole.

example exists in the world than the Totem Pole (stop 3). It is so narrow, it looks like it might topple over at any second. The protective cap of Shinarump conglomerate it undoubtedly wore like a helmet has long since tumbled away in one of the many landslides that shaped the spire. Few towers achieve such breathtaking slimness combined with such prodigious height—the rocks that form them are usually too weak to support such weight. Even now, erosion is laying the groundwork for the next landslide, which just might be the Totem Pole's last. The impending demise of the Totem Pole and its neighboring spires, the Yei Bi Chei, named for Navajo religious dancers, will erase the last traces of the mesa that once stood here, reducing this patch of ground to another nondescript bit of low-lying valley.

The carving of the Totem Pole and the rest of Monument Valley was indeed a monumental achievement. If not for the exact sequence of sedimentary rock layers and their fortuitous upwarping, erosion might never have had the chance to sculpt this iconic landscape of the American West.

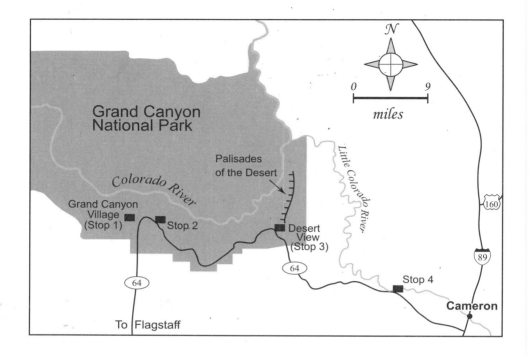

◆ GETTING THERE

This excursion begins at the south rim of Grand Canyon National Park, reached by traveling north from Flagstaff on U.S. 180 to Arizona 64. Once inside the park, follow signs to Grand Canyon Village and park as close to Bright Angel Lodge as possible. Walk west 200 yards from the lodge to the Hermits Rest bus stop. Pick up the Rim Trail at a set of five stone steps just west of the bus stop. Follow this trail west for 300 yards to stop 1, the signed Interfaith Worship Area, also called Worship Point. To reach stop 2, the South Kaibab trailhead, return to the Hermits Rest bus stop and catch a Village Route bus (blue line) to the Canyon View Information Plaza; then a Kaibab Trail Route (green line) bus to the trailhead (day-use visitors may not drive to the trailhead). This hike—about 2 miles round-trip with a 600-foot elevation change—takes about an hour and a half. Stop 3 is the Desert View overlook; drive east from Grand Canyon Village on Arizona 64 for 25 miles. To reach stop 4, the viewpoint of the Little Colorado River gorge, continue east on Arizona 64 another 23 miles and turn left 0.8 mile past milepost 285 at a sign marked "Scenic View." Park and walk 200 yards, through the Navajo crafts vending stands, to the overlook.

Journey to an Ancient Shoreline
THE GRAND CANYON AND THE GORGE OF THE LITTLE COLORADO RIVER

The Grand Canyon is one of the geologic wonders of the world. Hewn by nature from the raw material of brightly colored, horizontal bands of rock stretching to the horizon, the canyon is breathtaking in its immensity and visual complexity. The canyon's vibrant layers are composed of sedimentary rocks, and as with any great sculpture, inherent variations in this stone control the visual impact of the final work of art. In this case, main themes repeat themselves from viewpoint to viewpoint, but the intrinsic variations both within and between each sedimentary layer imbue each vista with unique and powerful qualities.

Sedimentary rocks are like history books; their pages reveal the geography, climate, and ecology of a region hundreds of millions of years ago. Etched on the canyon's steep walls are vivid tales of tropical seas come

formation	age (million years ago)
Kaibab formation — limestone (marine)	270
Toroweap formation — limestone and evaporite (tidal flats)	273
Coconino sandstone — desert sand dunes	275
Hermit formation — river floodplain muds	280
lower canyon layers	

The four highest rock layers in the Grand Canyon

93

and gone, mudflats stretching to the horizon, and a desert filled with sand dunes 1,000 feet high. Every Grand Canyon layer tells its own fascinating story, but this vignette focuses on just the four uppermost layers and their tales. These layers are the Kaibab formation, Toroweap formation, Coconino sandstone, and Hermit formation. We will examine the rocks' inherent variations in composition and properties, and how these differences contribute to the form of today's canyon. Then we'll take a journey back in time to walk the ancient landscapes where these sediments were laid down.

From the vista point of stop 1, look at the wall of rock to the left (northwest). The top 300 feet consist of layered, gray limestone cliffs belonging to the Kaibab formation. This rock forms the entire rim of the Grand Canyon, including beneath your feet, so you can easily examine it for yourself. A few small slopes interrupting the cliff face reveal the presence of muddier layers within the limestone. Mudstone is crumbly and erodes quickly, forming gentle slopes. Undercut cliffs above collapse onto the slopes. This process, called scarp retreat, is the primary mechanism by which the Grand Canyon has widened to its present immensity. The Kaibab ends at the break in the slope at the base of the cliff.

The upper layers of the Grand Canyon as seen from Worship Point

Below the gray Kaibab cliff, a well-defined slope has formed in the next layer, the Toroweap formation. The slope is green with vegetation. Although you won't be able to examine it closely until the next stop, its relatively gentle topography is a strong clue that it consists mainly of soft rocks, including mudstone. One exception to this is the discontinuous, gray-brown cliff band disrupting the gentle slope about two-thirds of the way down. This layer is hard limestone reminiscent of what we saw in the Kaibab, so it forms a resistant cliff instead of a gentle slope.

Below the Toroweap slope is a dramatic blond cliff of Coconino sandstone, whose boundaries are so easy to spot, you can easily trace them for miles along the North Rim, 10 miles distant. Like the Kaibab, the bottom of the Coconino formation coincides with the bottom of the cliff, which ends at another vegetated bench composed of bright red soil. This bench lies in the soft Hermit formation, one of the canyon's most prominent mudstone layers.

The horizontal nature of the Grand Canyon's rock layers has profoundly influenced the way these rocks have responded to erosion and hence has dictated the broad outline of the canyon's topography. Sheer canyons form only when a river is cutting down into a mass of resistant rock. As the Colorado River carved the Grand Canyon, it first encountered the resistant Kaibab limestone, which was strong enough to support a narrow canyon. But as the river continued to eat its way down through the layer cake, it soon met the Toroweap formation. The river hungrily devoured these soft rocks, undercutting the steep limestone cliffs above and causing them to collapse in a series of landslides—scarp retreat at work. As the Colorado continued to cut down, it alternately encountered hard and soft sedimentary layers. At every resistant layer, the river carved a narrow canyon immediately above its banks, and at every soft layer, another wave of scarp retreat began. This process has led to the canyon's stair-stepped appearance.

Now that we can identify the canyon's uppermost layers, we can search for clues to reconstruct the landscapes in which these sediments were deposited. First, we'll compare the layers vertically to see how conditions changed with time at a single location. Then, at Worship Point and three other stops, we'll use lateral changes in each layer to reconstruct a detailed picture of what the geography of northern Arizona looked like 280 to 270 million years ago.

If you examine the Kaibab formation closely at Worship Point, you will see that the gray limestone is noticeably lumpy. These lumps are made of a smooth, very hard substance called chert, a form of pure silica with microscopically small crystals. Look at enough of these chert lumps, or nodules, as they are called, and you will find a fossil seashell encased in

A brachiopod fossil at Worship Point

one. Most of the nodules contain a fossil, and once you train your eye, you'll be able to find dozens of them. Most of the fossils resemble clam or oyster shells, but they actually belong to a group of organisms called brachiopods. Other fossil species here are rarer but not too hard to spot if you examine enough nodules. Horn corals look like tiny cornucopia, and two varieties of bryozoans look, respectively, like branching tree twigs and coarse mosquito netting. Crinoids, commonly known as sea lilies, have cylindrical, ribbed stalks. When these organisms died, the stalks broke into pieces at each rib, forming button-shaped disks that settled flat on the seafloor. Most of the crinoids you will see here are in the form of these cross-sectional disks, which resemble small, spoked wheels. Finally, there are sponges, whose contorted patterns resemble a brain or a pile of spaghetti.

    The presence of these fossils speaks volumes about the ancient environment in which the Kaibab limestone accumulated. Since all of the fossils are the remains of marine organisms, it is easy to conclude that, despite its current lofty elevation, this spot was submerged under an ocean 270 million years ago. Through comparison of these fossils with their nearest living relatives, we can further conclude that these species lived in the relatively shallow waters of the continental shelf, and that

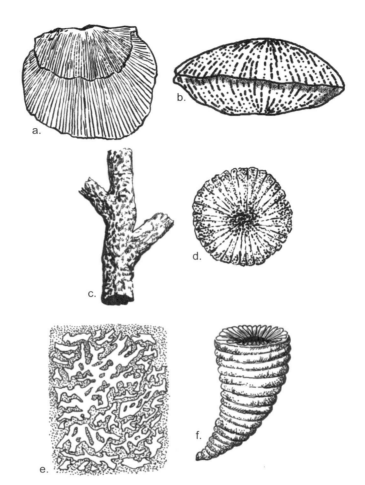

Fossils found in the Kaibab formation at Worship Point: a) brachiopod viewed from above; b) brachiopod viewed edge-on; c) bryozoan; d) segment of crinoid stem; e) sponge; f) horn coral —Dona Abbott

they preferred well-stirred, open marine conditions of normal salinity. In addition, a small amount of gritty, brown sandstone mixed with the limestone here suggests that this spot was not too far from shore, with its terrestrial supply of sandy sediment.

But why are virtually all of the fossils in this limestone encased in nodules of chert, a different type of rock? The answer begins with the sponges that inhabited the sea in which the Kaibab was deposited. Unlike most other marine invertebrates, sponges don't secrete a shell of calcium carbonate—the material that forms limestone. They do, however, possess small, hard internal structures called spicules, which, like bones, help these gelatinous creatures maintain their shape. The spicules are made of opaline silica, a type of silica that, importantly, has a bit of water in its chemical structure.

When a sponge dies and decays, the spicules fall out of it like needles from a rotting pincushion. So many sponges inhabited the Kaibabian sea that its floor was, quite literally, littered with spicules. The opaline form of silica is quite soluble, so once the spicules were buried under younger layers of sediment, they began to dissolve, saturating the water with silica. The silica-saturated water then percolated through the limestone sediment and the fossils embedded there. Many of these still contained small amounts of undecayed organic material, which altered chemical conditions around the fossils and allowed chert to precipitate. Unlike opaline silica, chert has no water in its crystal structure. This seemingly small difference is crucial in the formation of the nodules and in the lumpy nature of the Kaibab limestone. Unlike opaline silica, chert is tough as nails and resists erosion even better than limestone. The protruding lumps of chert nodules all around you stay put while the limestone erodes down around them.

The Kaibab formation forms the Grand Canyon's rim along its entire 230-mile length, and these abundant chert nodules help explain why. Many other sedimentary rock layers used to overlie the modern land surface in the Grand Canyon area, but millions of years of erosion stripped them away. However, when the forces of erosion hurled themselves at the hard Kaibab limestone and its armor of chert, they were resoundingly defeated. Only the powerful Colorado River has been able to breach this nearly invincible layer. The Kaibab thus forms the land surface everywhere except where the river sliced through, forming the Grand Canyon.

Now that we have a handle on the characteristics of the Kaibab formation at Grand Canyon Village, let's take a short hike on the South Kaibab Trail (stop 2) to examine the underlying layers. The trail begins with a dizzying series of switchbacks through the Kaibab formation, which hasn't changed appreciably from what you saw at Worship Point. Fossils like those you saw at the first stop are common here as well, especially in the rocks at the fifth switchback, but they aren't as easy to spot. Look for them in the resistant white and yellow chert nodules and bands.

When you reach the bottom of the switchbacks and the trail gradient eases noticeably on a long, descending traverse about 0.3 mile from the trailhead, you have arrived at the underlying Toroweap formation. It is obvious that, as we noticed at stop 1, the Toroweap here is a soft, easily eroded unit. Because the rock in most places is mantled with soil and vegetation, you need to walk another 0.2 mile, about halfway across the long, straight traverse, to examine it in detail. Here you see, to your right, a 1-foot-thick, white limestone layer that is buckled into an arch just above the trail. Below the limestone, a jumbled mass of soft, chalky, fine-grained rock sports a range of colors, including red, brown,

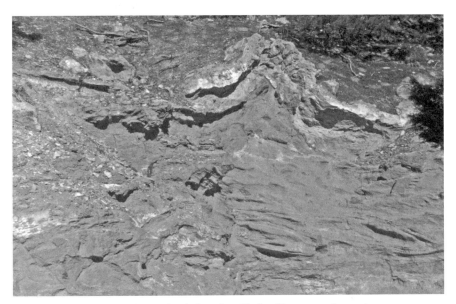

Upward flow of salt has buckled a limestone layer in the Toroweap formation along the South Kaibab Trail.

green, and white. As we surmised at the first stop, the erosion-prone, slope-forming rock consists mainly of mudstone, along with another soft rock, evaporite—accumulations of salt laid down as a body of salty water was evaporating. These are the dominant materials comprising the Toroweap formation here. It is the abundance of evaporites that led to the buckling of this limestone layer. Evaporites, including halite (table salt) and gypsum (used to make drywall), are less dense than other common rock-forming minerals. This means that when they are buried beneath other sediment, they become buoyant, flowing upward wherever overlying layers have a weakness and buckling these layers in the process, as you see here. The salt rises for the same reason helium balloons do: it is lighter and more buoyant than its surroundings. It rises until it reaches equilibrium.

The abundance of mudstone and evaporite in the Toroweap formation provides important clues about its depositional environment. The evaporite is the most diagnostic. As the name implies, evaporites form in environments where salty water is evaporating, such as an arid tidal flat or a desert salt pan like those in the bottom of Death Valley today. Marine fossils have been found in the Toroweap's limestone layers, indicating a marine origin for the formation and thus favoring the tidal-flat option.

About 273 million years ago, the Grand Canyon was a vast, arid tidal flat reminiscent of those common today around the Persian Gulf. The buckled limestone layer you see here was laid down during a slight rise in sea level, which briefly submerged this site. The same thick limestone layer we noted from Worship Point lies another 0.2 mile down the trail, just above the contact with the underlying Coconino sandstone. This limestone was deposited during a longer-lasting rise in sea level. Its presence helps confirm that the Toroweap is indeed a marine tidal-flat deposit.

As you continue to descend, you reach the obvious contact with the Coconino sandstone about 0.8 mile from the rim. The Coconino consists of blond, fine-grained sandstone arranged in distinctive steep-angled layers called crossbeds. Crossbeds are signs of deposition in a desert sand dune environment (more in vignette 4). Sandstones of similar age, with nearly identical characteristics, have been found all the way from here to Colorado, indicating that a vast desert covered the American Southwest 275 million years ago.

A promontory just below the contact (about a mile from the trailhead) provides a grand vista into the canyon. From here you can look down on the flat, red bench of Cedar Ridge below (with a large solar toilet built on it). This bench has been carved into the easily eroded Hermit formation, whose fine-grained mudstone was deposited on an expansive river floodplain 280 million years ago (more about these rivers in vignette 13). If you choose to hike an additional 0.5 mile down to Cedar Ridge, you can examine fossil impressions of ferns that grew along the banks of those 280 million-year-old rivers. To see the fossils, walk west from the toilets to a rock enclosure, where specimens are under glass.

All four of the Grand Canyon's uppermost layers were laid down as sediments during Permian time, before the dinosaurs or any mammals had evolved. By comparing and contrasting the characteristics we have observed in these layers, we can assemble an amazingly clear picture of how dramatically the topography and climatic conditions have changed in this spot.

What is now northern Arizona was, during late Permian time, a vast, low-lying plain alternately inundated by shallow seas and exposed as flat coastal plains. Two factors mainly controlled the type of sediment deposited on this plain: sea level, which dictated the position of the shoreline and the depth of the sea; and regional climate, which repeatedly swung between wetter and drier throughout the period.

Around 280 million years ago the Grand Canyon enjoyed a semi-arid climate similar to that of today. At the time, sluggish, meandering streams crisscrossed a vast, flat river floodplain, much as the modern Mississippi lazes its way across low-lying Louisiana. These rivers deposited the

Groups of slanting crossbeds in the Coconino sandstone are visible on the rock face (right). *Horizontal bounding surfaces—lines separating sets of crossbeds—mark where one sand dune planed off the top of the dune that preceded it.*

Hermit formation. As the climate became progressively more arid, the rivers dried up, and sand dunes up to 1,000 feet high began to blow across the plain in a huge, Sahara-like desert around 275 million years ago. The dunes' legacy is the Coconino sandstone. Just 2 million years later, sea level was on the rise, and seawater crept in to cover the area. The climate was still extremely arid, so the tidal flat that occupied the Grand Canyon area endured extensive evaporation, leaving behind the salts and mud of the Toroweap formation. Rising sea level eventually inundated the tidal flat, and by 270 million years ago, a warm, shallow sea rife with marine organisms covered the area. The Kaibab formation preserves some of these creatures.

This vertical comparison allows us to observe how depositional conditions changed with time. We can also reconstruct the geography of the area at one snapshot in time—known as its paleogeography—by observing variations in rock type or fossils within individual layers. At the wonderful overlook at Desert View (stop 3), take a look at the Kaibab formation, which once again forms the bedrock at the rim. There are a few noteworthy differences between the Kaibab here and at stops 1 and 2. While still reasonably abundant, lumpy chert nodules are less common here than they were farther west. Fossils can still be found in the nodules, although there are fewer than you saw by Grand Canyon Village. The one feature in greater abundance here is sandstone. At Worship Point, lens-shaped deposits of sandstone were few and far between, whereas here, sandstone lenses comprise about 30 percent of the rock. Wander around a wide swath of the rim and get a sense of the relative abundance of sandstone.

What do these differences tell us about the area's paleogeography? As we head east, the increasing sandstone content is most easily explained if we are getting closer to the ancient shoreline, the ultimate source of sand. This discovery further suggests that during Permian time, the land must have sloped down to the west, so that when sea level rose, the water would have encroached from that direction.

To scrutinize the layers under the Kaibab, climb to the top of the Desert View tower. This is one of the few places on the rim where you can get an unobstructed view of the Colorado River. The imposing rock wall to the right of the river is called the Palisades of the Desert, a nearly vertical escarpment rising 4,000 feet from river to rim. Clearly, the mudstones that formed gentle slopes in the stair-step topography farther west are less prominent here. As you try to pick out each layer, you may wonder whether the Toroweap and Hermit formations, whose gentle slopes are so obvious farther west, have disappeared. Although you can still spot a small break in slope that marks the Kaibab-Toroweap boundary about 300 feet below the rim, the Toroweap contains much more cliff-forming sandstone here than it did to the west. The crossbedded Coconino cliff is still easy to spot, with the red Hermit formation still present just below the sandstone. The Hermit is so thin, however, that it is now difficult to distinguish from the equally vivid red cliffs of the underlying Supai group. The Kaibab, it seems, is not the only layer that has changed in character.

Keep these variations in mind as you head to stop 4, an overlook of the obscure gorge of the Little Colorado River. What this gorge lacks in depth and immensity, it more than makes up for with its walls' dramatic 900-foot plunge. Anywhere else, this spectacle would be a widely acclaimed tourist attraction, but here, in the shadow of the Grand Canyon, most people simply pass it by.

As at the previous stops, the rim of the gorge at stop 4 belongs to the Kaibab formation. Plenty of the formation is on display along the short stroll from the parking lot to the overlook. The rock is mostly a sandy limestone, suggesting that we are drawing still closer to the shoreline that existed during Kaibab deposition. In addition, chert nodules and their constituent fossils are nearly absent. Apparently, few sponges inhabited this portion of the Kaibabian sea.

Detailed studies by paleontologists who have scoured this region for fossils bear these observations out. There are fewer fossils in the Kaibab here and to the east. More importantly, the fossils that are present here are predominantly mollusks of a variety that tolerates elevated salt concentrations. You can see fossils of a few button-sized mollusks in the rock ledge just behind the railing at the main overlook. Look inside several

The narrow gorge of the Little Colorado River cuts through the bedded Kaibab formation and the darker Toroweap and Coconino formations below (indistinguishable from one another here).

pits in the rock to see them. Such evidence strongly suggests that this area was a restricted, near-shore arm of the sea that lay in an arid environment. Intense evaporation elevated the salinity in this bay, forcing out all but the hardiest species.

At the impressive overlook, if you try to trace our four layers down the walls of the gorge, you will realize that they don't look completely familiar. The Kaibab displays marked horizontal bedding from the rim to about the 300-foot level, as we have observed before. But the gentle slope we saw in the Toroweap at stops 1 and 2, attributed to the soft rocks in the formation in those locations, is entirely absent. The sheer nature of the gorge walls indicates that, here, the Toroweap is made of more resistant rock types than the mudstones and evaporites of stops 1 and 2. The crossbeds faintly visible in the sandy rock immediately below the Kaibab are reminiscent of those you saw earlier in the Coconino sandstone. The Toroweap formation is present here, but its character has changed dramatically to a crossbedded sandstone nearly indistinguishable from the Coconino. This means that the sand dunes that occupied the Grand Canyon region during the Coconino's deposition persisted at this location throughout the time the Toroweap was laid down. This spot wasn't inundated by the eastward-creeping sea until the time of Kaibab deposition, about 3 million years later. The shoreline of the sea during Toroweap time lay at about Desert View overlook, stop 3, where the Toroweap mudstones and evaporites thin to a sliver and wink out. The sand dunes frozen in stone in the gorge below you marched to the shore of that sea. The floor of the Little Colorado River gorge lies within the Coconino sandstone, so we are unable to observe the Hermit formation here.

Following changes in each of the Grand Canyon's top four layers from west to east allows us to relive the area's changing geography and surface conditions. We can even trace an ancient shoreline from 280 to 270 million years ago.

The consistent, clearly terrestrial nature of the Hermit formation river deposits and the Coconino sandstone dunes along our transect shows that while these layers were being deposited 280 to 275 million years ago, the shoreline lay in western Arizona. However, by 273 million years ago, when the Toroweap was being deposited, rising sea level had shifted the shoreline eastward. The mixture of terrestrial sand and softer evaporite in the Toroweap at the Desert View overlook indicates that the shoreline lay near there, with tidal flats present to the west along the South Kaibab Trail. A trip west from Grand Canyon Village at that time would have taken you from the tidal flats into open ocean, where limestone was deposited. That limestone is an abundant component of the Toroweap in the western Grand Canyon.

During deposition of the Kaibab formation, as the sea continued to encroach from the west, each of these environments was gradually inundated, and the shoreline shifted even farther east. The preponderance of limestone in the Kaibab at all four stops shows that the entire area was completely submerged by about 270 million years ago. The increasing abundance of shore-derived sandstone at Desert View overlook and the Little Colorado viewpoint indicates that the shoreline lay not terribly far to the east of this gorge. The fact that the Kaibab limestone disappears entirely about 70 miles farther east, along a line running north from the town of Holbrook, suggests that the shoreline lay there during Kaibab deposition.

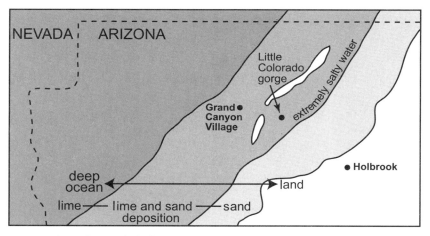

When the Kaibab formation was deposited 270 million years ago, a warm, shallow sea had encroached from the west, covering northern Arizona. The shoreline was farther east, near Holbrook.

The Grand Canyon is often hailed as the planet's preeminent time machine, taking you layer by layer back through the deep history of the earth. Now you have experienced it. You have strolled an ancient shoreline without leaving the landlocked Southwest.

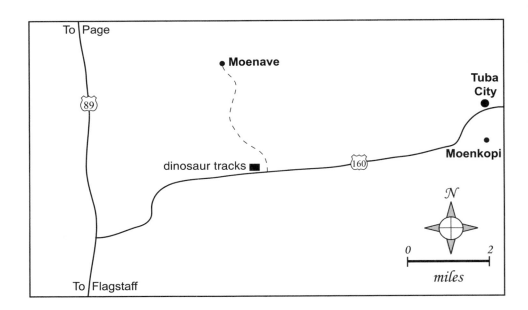

◈ GETTING THERE

The track site is in northeastern Arizona 5.4 miles west of Tuba City and 5 miles east of the junction of U.S. 160 and U.S. 89. On U.S. 160, 0.3 mile east of milepost 316, turn north on the clearly marked dirt road that leads to the village of Moenave. Drive just 0.2 mile and stop in a wide parking area on the left next to a series of sun shelters. The tracks lie immediately behind the shelters. Guides in the shelters offer tours free of charge, but tips are appreciated.

Dilophosaurus Dancehall
MOENAVE DINOSAUR TRACKS

A group of dinosaurs waltzed across the area near Moenave some 200 million years ago and left an outstanding collection of over three hundred footprints. Just a few steps from the road, the tracks are remarkably preserved in stone slabs of the Jurassic-aged Kayenta formation. Some of the tracks are mere smudges, but others preserve exquisite detail, right down to razor-sharp claw marks. Single footprints, trackways half a dozen prints long, and an actual piece of broken-off claw reveal moments in these prehistoric creatures' lives as they tramped across a muddy plain.

The Moenave prints crisscross an area the size of a football field, stretching northwest from the parking area to a diminutive canyon. Most of the tracks are large, up to 12 inches long, and tridactyl, or three-toed. The three toes are characteristic of meat-eating dinosaurs. No undisputed similar prints exist from the preceding time period, the Triassic, so these tracks provide evidence of some of the world's first large meat-eating dinosaurs. In addition, because similar prints have now been

One of the many Eubrontes *trackways at Moenave*

found across the globe, from China to South Africa and Europe to the American Southwest, the Moenave tracks have far-reaching implications for understanding both dinosaur evolution and the last days of an ancient supercontinent.

To truly appreciate the global significance of the tracks at Moenave, however, a quick trip east to New England is in order. It was there, in the 1830s, that footprints of this type were first studied—and the science of dinosaur tracking was born.

In 1836, Reverend Edward Hitchcock began a detailed study of an extraordinary line of tridactyl footprints fossilized in a slab of Connecticut stone. The birdlike tracks were enormous, up to 15 inches across. Reverend Hitchcock named them *Eubrontes giganteus*, which means "enormous true thunder." Hitchcock's use of a Latin name inaugurated the practice that continues today of assigning genus and species to a set of tracks regardless of which species actually made them. Paleontologists do the same with fossilized skeletons. In the meantime, people nicknamed Hitchcock's assumed avian trackmaker "Noah's Raven."

Like everyone else, Hitchcock assumed a large bird had made the striking tracks. After all, he began his investigations before the concept of dinosaurs had emerged; the term "dinosaur" was not even coined until 1842. But over the next thirty years, as he studied the thousands of additional tracks unearthed across the eastern seaboard, Hitchcock came to realize that dinosaurs had made the prints. From the prints, he inferred many interesting details about the trackmakers' anatomy and habits, just as skilled hunters do when pursuing living animals. Tracks tell scientists much about an organism that they can't learn from its fossilized bones, and vice versa. Tracks and bones together provide complementary information, but only rarely are the two unambiguously tied together. The identity of the *Eubrontes* trackmaker remained unknown until 1942, when the Moenave track site achieved worldwide recognition.

That summer, Sam Welles, a University of California paleontologist, spent a field season in northern Arizona. The tracks near the village of Moenave were already known by that time, and the concept of dinosaurs was well accepted, but which creature had made the Moenave tracks remained a mystery. Just as Welles was preparing to leave, he heard of a dinosaur skeleton discovered two years earlier near Moenave by local resident Jesse Williams. Welles rushed to the site and found the remains of three large dinosaurs tightly clustered together in rock of the Kayenta formation just 400 yards from the *Eubrontes* tracks. Welles named the dinosaur *Dilophosaurus* (dill-OFF-uh-SOAR-us), which means "double-crested reptile," and noted its foot was perfectly suited for making the *Eubrontes* tracks.

This cast of a Moenave Dilophosaurus *skeleton is on display in the Navajo Nation Museum in Window Rock, Arizona.*

If a fossil discovery comparable to the Moenave dilophosaurs were made today, the skeletons would likely be kept in place, protected, and made into a major tourist site. You would be able to examine the bones of *Dilophosaurus* right next to its tracks. However, when Sam Welles was working, preserving the bones in place was not even a consideration; the singular finds were hastily excavated and carted back to California. These fossils and a fourth, better-preserved skeleton Welles found nearby in 1964 are all on display in Berkeley at the University of California Museum of Paleontology. But you don't have to go all the way to California to marvel at *Dilophosaurus*'s impressive size and power; Flagstaff's fabulous Museum of Northern Arizona displays a reproduction of the best Moenave skeleton, and a fossil cast is featured in the Navajo Nation Museum gift shop in Window Rock, Arizona.

The Moenave site was and remains a particularly significant find. The early Jurassic was a pivotal time in dinosaur evolution. It saw the animals' initial rise to prominence, and quickly thereafter, dominance in terrestrial ecosystems around the globe. Maddeningly, however, the early Jurassic dinosaur fossil record is skimpier than that of most other time periods. Welles's discovery of prints adjacent to the remains of the likely trackmaker filled a major gap in our knowledge of dinosaur life during this critical time.

As you head northwest from the parking lot, note how all the prints in the first trackways you encounter (immediately behind the guides' wooden booths) are the same size. These tracks were made by dinosaurs walking erect, so you only see impressions of their hind feet. In contrast, another trackway about 30 yards north of the parking area features both large and small prints in succession. These were likely made by the larger hind and smaller front feet of a slow-moving dilophosaur who periodically came down into a crouch, instead of walking erect as members of the species did when they were in a hurry.

A quick glance around the site shows a lot of variability in the tracks' degrees of preservation. In truth, we are lucky to see any at all. Our ability to study dinosaurs and other extinct creatures depends entirely on nature for the conditions necessary to both preserve and reexpose such delicate relics. As difficult as track preservation is, preserving bones is even more

Where tracks are well preserved, as these are, even claw marks are visible. In this trackway, the larger print is from the hind foot; the smaller print was likely made by the animal's forefoot as it bent down into a crouch on the previous step. Quarter for scale.

Preserved for 200 million years, this piece of claw likely belonged to a dilophosaur.

difficult. Intact skeletons like those Jesse Williams found are extremely rare; only one other dilophosaur skeleton has ever been found.

Dinosaur tracks are more common than fossils. Still, only a tiny fraction of tracks escape nature's eraser. Tracks like these can be preserved in two different ways. The first, the "cover-up" method, occurs when an animal makes a print in sediment that is moist but not completely saturated. If the print dries out and hardens, then is quickly buried by an influx of sediment—say, from a flood—it stands a good chance of being preserved. In a second scenario, known as the "underprint," an animal is heavy enough to make an impression not just on the earth's surface but also in underlying layers. Already buried, these deeper imprints have a greater chance of long-term preservation.

Even if tracks are preserved, they may never be reexposed. The sediments hosting the prints must be buried deeply enough to transform into rock, but not so deeply that they are metamorphosed. Erosion must then take over, stripping off enough overlying rock to reveal the host layer, as well as removing sand grains that originally filled the fragile prints. Clearly, with such exacting conditions, most prints never make it, let alone come to light.

Exploring around the tracks at the Moenave site, you will see that many of them display fiercely pointed claw marks. Such exquisite detail is often lost in underprints, so it's likely the cover-up method saved these tracks. About 70 yards northwest of the parking area, you can even see a broken-off piece of claw preserved in the rock—an obligatory stop on all the guided tours.

When the dilophosaurs danced around this area, it looked very different from what you see today. During Jurassic time, a range of tall

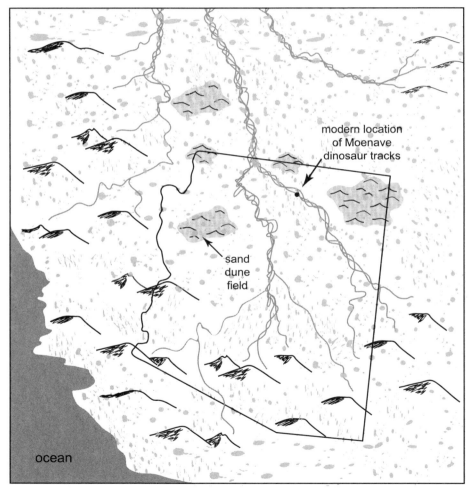

Northern Arizona during early Jurassic time was a low, semi-arid, flat plain crisscrossed by braided rivers flowing from volcanic mountains in southern Arizona.

mountains in southern Arizona fed a family of braided rivers that spilled northward across a series of flatlands on their way toward the sea in Utah. Shallow lakes dotted this landscape, and fossilized bits of wood indicate that abundant vegetation lined the lakes and river floodplains. Sandstone and siltstone layers deposited by these rivers in flood likely made the cover-up of these Moenave tracks and skeletons possible.

The Moenave skeletons reveal that *Dilophosaurus* was the biggest, fiercest carnivore of the time. A full-grown dilophosaur stood about 7 feet tall, stretched 18 feet long, and tipped the scales at 600 to 700 pounds. The function of its namesake double head-crest is not known. Because it would add to *Dilophosaurus*'s already intimidating appearance, the crest may have served to frighten rivals or attract potential mates during courtship.

Dilophosaurus achieved a measure of fame as the poison-spitting dinosaur in the movie *Jurassic Park*. There is no scientific evidence that *Dilophosaurus* was actually equipped with such poison, but it was amply armed for combat with rows of jagged teeth and, on its feet and forelimbs, razor-sharp claws like the one preserved here. Based on measurements of the spacing between prints on some of the longer *Eubrontes* trackways, scientists estimate that *Dilophosaurus* was very swift, capable of running about 15 miles per hour over short distances.

An artist's interpretation of Dilophosaurus, *with its distinctive double crest* —Dona Abbott

Traditionally, paleontologists have assumed that carnivorous dinosaurs preyed on their plant-eating cousins. However, for the dilophosaurs, facts on the ground here don't fit that assumption. In Jurassic time, there were a number of plant-eaters on which dilophosaurs could have preyed, including the 8-foot-long *Massospondylus*. However, rock layers in this area from which *Massospondylus* fossils have been unearthed have characteristics indicative of arid, dune-covered highlands. The tracks of plant-eaters such as *Massospondylus* have been rare or completely absent from places where *Eubrontes* tracks are found. This is true at Moenave, where *Eubrontes* tracks dominate, and rock features invariably indicate a wet bottomland environment like a lakeshore or river floodplain.

Also casting doubt on the idea that *Dilophosaurus* was a fearsome predator, paleontologists studying the skeletons noted its tooth-studded snout bone was too delicately attached to the skull to withstand the forces generated by a powerful chomp on prey. They speculated that *Dilophosaurus* might have been more of a scavenger than a primary predator, pecking away at its meals like a vulture.

What, then, did dilophosaurs eat? Paleontologists were stumped until swim tracks were discovered in rock at Zion National Park and the Johnson Farm Dinosaur Discovery Site near St. George, Utah. The swim tracks consist of hind footprints adjacent to one another, marking where the animal had pushed off from the lake or river bottom with both legs. As you can see at Moenave, walking tracks show the alternation of hind footprints as the animal placed one foot in front of the other.

Once the paleontologists found the swim tracks, they quickly revisited the Moenave skeletons. Was their anatomy suited to eating fish? Examination of the delicate jaw of *Dilophosaurus* has shown that it was ideal for catching and eating fish, and that it is similar to jaws of known fish-eating species from the past and present. The jaw's slender profile would have trapped a minimum of water during each aquatic chomp. With nostrils far up its snout, *Dilophosaurus* would have been able to rest its jaw in the water for long, still periods, like the modern crocodile, waiting for passing fish; and its front teeth formed a large, deadly sieve for trapping fish. Abundant fish scales and skeletal material from both Zion and Johnson Farm indicate that a diverse and rich fish fauna inhabited these lakes, further bolstering this argument.

There is also evidence that dilophosaurs likely hunted in groups. Numerous parallel swim tracks at the St. George site suggest an entire herd swimming together against a mild current. Likewise, the large number of tracks at Moenave, though more randomly distributed, suggest that a number of dilophosaurs trekked through this area over a relatively short period of time. These revelations about *Dilophosaurus*'s diet and social behavior would not have been possible without the paired fossils and tracks here at Moenave.

Fascinating in and of themselves, the Moenave tracks and skeletons are also important evidence that the continents were joined together when the trackmakers lived. *Eubrontes* tracks appear in the fossil record suddenly and nearly simultaneously across the globe. From their skeletons, we know dilophosaurs could not fly, and while it appears they could

An artist's interpretation of the plant-eating dinosaur Massospondylus —Dona Abbott

swim, surely they couldn't have swum from China to North America. The only plausible way for a newly evolved land-based species to disperse so widely in such a short period of time would be to walk. Geologists see the presence of *Eubrontes* tracks on multiple continents as supporting evidence for the existence of the supercontinent Pangaea. As such, the tracks also support the theory of plate tectonics, in which the world's landmasses move across the globe, assembling and reassembling. Pangaea broke up not long after these tracks were made.

Understanding previous continental configurations is important for many reasons—for example, locating valuable resources such as oil, gas, and some minerals. But the story of Pangaea becomes still more germane when you consider that the 100 million years during which it existed were crucial for the development of life on our planet. How the continents were arranged undoubtedly played a role in how life evolved.

The evolution of dinosaurs is particularly fascinating because it is bookended by two of the biggest extinctions in the earth's history. The dawn of these creatures came in Triassic time, about 230 million years ago, when Pangaea was in its prime. The earliest dinosaur skeletons found in northern Arizona and around the world include that of a diminutive, bipedal creature named *Coelophysis*. But, as described in vignette

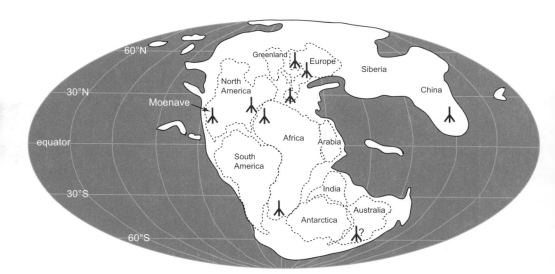

The configuration of the supercontinent Pangaea during early Jurassic time. The track symbol denotes locations where Eubrontes *tracks have been found. The Australian location is in older (Triassic) rocks, and it remains controversial whether they are truly* Eubrontes.

15, these small early dinosaurs were relatively insignificant in the late Triassic ecosystems they inhabited.

The fortunes of this lineage changed when descendants of *Coelophysis*, the forebears of Moenave's *Dilophosaurus*, survived the massive extinction at the end of the Triassic, an event in which over 50 percent of all the earth's species were expunged, just prior to the formation of these tracks. First discerned by a break in the marine fossil record, this extinction was one of the five largest in the history of the world, and its importance for evolution cannot be overstated. After such a profound die-off, new species tend to evolve rapidly, taking advantage of freed-up resources. Without these unclaimed resources, many of the organisms that repopulated the earth could never have developed, so creatures such as dinosaurs and perhaps even humans might never have evolved.

A controversial hypothesis about the causes of this end-of-Triassic extinction has lately gained scientific credence. The hypothesis posits that an asteroid impact triggered sudden and catastrophic climate change, causing the earth's ecosystems to collapse. Just such a collision is thought to have caused an even larger extinction event 65 million years ago, when an asteroid impact near Mexico's Yucatan peninsula began a chain of events that killed off a whopping 75 percent of all the species on earth, including the dinosaurs (vignette 14).

The many *Eubrontes* trackways around the world, including the ones here at Moenave, have assumed a central role in the debate about the end-of-Triassic extinction event and whether an asteroid caused it. Several lines of evidence dating from the Triassic-Jurassic boundary include a spike in the extraterrestrial element iridium at several locations around the world and a "shocked" form of quartz that forms only under extreme pressures. This evidence suggests an impact did indeed occur.

Also of significance, sediment laid down in Connecticut a mere 10,000 years after the Jurassic period began contains an explosion of *Eubrontes* tracks identical to the ones found here at Moenave and around the world. No conclusive *Eubrontes* or other large dinosaur tracks have been found in older, Triassic-aged rocks. Many paleontologists interpret this rapid, worldwide proliferation to mean that *Dilophosaurus* and other large dinosaurs evolved suddenly and spread quickly after a global catastrophe.

While the marine fossil record near the Triassic-Jurassic time boundary clearly indicates that this was a major extinction event, the sparse terrestrial fossil record is sometimes interpreted differently, causing some paleontologists to disagree. This latter group construes the terrestrial record as containing not one single catastrophic extinction but rather a series of smaller events over a period of several million years. Much of the ambiguity may be due to the fact that the marine record is much more plentiful than the terrestrial. Many more organisms flourished in the seas, and conditions there generally lend themselves to better fossil preservation than on land.

Giving tenuous support to the multiple-small-extinction hypothesis, paleontologists in Australia have found three poorly preserved footprints in Triassic-aged rocks from a Queensland coal mine that they have interpreted as *Eubrontes*. The Australian tracks are truly enormous, over 19 inches long—about 20 percent larger than the tracks at Moenave. The Australian paleontologists conclude from these tracks that large dilophosaurs were thriving in southern Pangaea during the late Triassic; the sudden appearance of their tracks in Arizona and New England in the early Jurassic would be due to their progressive migration across Pangaea, not to the rapid evolution of new species.

The Australian find would indicate that *Dilophosaurus* and possibly other large dinosaurs already existed before the extinction. Catastrophic events like asteroid impacts are believed to take a particularly heavy toll on especially large creatures like dilophosaurs by eliminating their food sources; evidence that they survived the event would cast doubt on the asteroid impact hypothesis. Other paleontologists disagree completely with this interpretation; they believe the Australian tracks belong to a different genus, probably not a dinosaur. Similar large non-dinosaurian tracks have been found in North Carolina. For now, many paleontologists believe identification of these three prints is tentative. They prefer to reserve judgment regarding the existence of Triassic dilophosaurs until skeletal material or more definitive *Eubrontes* tracks are discovered.

It is frustrating to have no definitive answers, and because we are entirely dependent on nature to preserve and unearth clues, it will likely be a while before this puzzle is finally solved. However, Moenave's remarkable tracks remind us how much we have learned in less than two centuries, since Reverend Hitchcock. The discovery of *Dilophosaurus* skeletons next to tracks here at Moenave ushered in an era of rapidly accumulating scientific studies, altering our image of the trackmakers from Noah's Raven to a herd of dilophosaurs doing water ballet.

TRACKING THE TRUTH

Dinosaurs are not always portrayed accurately in film or literature. Likewise, not all of the "feature attractions" that guides may show you at Moenave pass scientific muster. Noteworthy among these are the purported dinosaur rib cage—woe to any animal with such a lopsided set of bones—and the *Tyrannosaurus Rex* tracks. *T. Rex* didn't evolve until late Cretaceous time, over 100 million years after the sand and mud of the Kayenta formation—and these tracks—were laid down.

Some guides claim that curious hemispherical rocks littering the ground near the road are coprolites, or petrified dinosaur feces. Coprolites are fairly common in the fossil record, and these unusual rocks do resemble oversized cow pies. But no scientific study has been undertaken that supports this claim. You may also hear that a distinctly different, circular rock is a dinosaur egg. Dinosaur eggs have been found elsewhere in the world, but again, this particular one has not been studied scientifically. A much less exciting interpretation is that both the "egg" and the "coprolites" are simple concretions formed by mineral-charged water percolating through rocks. Erosion of the softer rock around a concretion exposes it, revealing its smooth, rounded surface.

As intriguing as coprolites and dinosaur eggs would be, the dinosaur tracks at Moenave and the detailed reconstructions of dinosaur behavior patterns that scientists have made from them are no less extraordinary. The stories are entertaining, but the truth makes an even better tale.

Guides at Moenave may tell you these curious oval mounds are coprolites—petrified dinosaur feces.

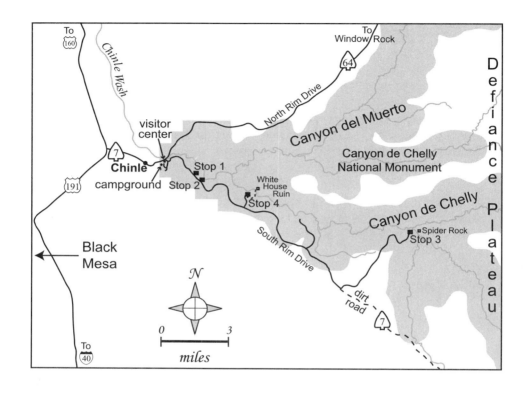

❖ GETTING THERE

Canyon de Chelly National Monument is in northeastern Arizona, just east of Chinle. From U.S. 191, go east on Navajo Route 7 for 2.7 miles to the monument's visitor center. To reach stop 1 at Tunnel Canyon Overlook, continue east another 1.9 miles on Navajo Route 7, which becomes South Rim Drive. Stop 2 is at Tsegi Overlook, 0.3 mile farther east on Route 7. Stop 3, Spider Rock Overlook, is 13.5 miles farther, at the end of the paved road. To reach stop 4 at White House Overlook, backtrack west 10.5 miles along Navajo Route 7 to the White House turnoff. Drive the 0.6-mile access road to the parking area and walk 50 yards northeast to the overlook. The trail to the ruin passes through a marked gap in the railing to the right (south) and continues across the slickrock. The moderate 1.25-mile trail descends 500 vertical feet. Monument visitors without a guide are allowed only at the overlooks and on the trail to White House Ruin.

9

Beauty All around Me, with It I Wander
CANYON DE CHELLY NATIONAL MONUMENT

Northern Arizona is a land of canyons. From the overwhelming majesty of the Grand Canyon (vignette 7) to stunning slots like Antelope Canyon (vignette 4), northern Arizona has more than its fair share. Compared to this first-class lineup, Canyon de Chelly (pronounced duh-SHAY) lacks superlatives. It's not the deepest canyon in Arizona. Many are narrower, and a host are longer. But despite its lack of noteworthy statistics, Canyon de Chelly exerts a hold on the mind and a tug on the emotions that easily rivals the spells cast by gorges with more impressive dimensions.

Maybe it is the deep human history that permeates the place, or maybe it is the elegant simplicity of the canyon's lines. It may be due to the fact that many people still call it home, going about their daily lives

The simple beauty of Canyon de Chelly

amidst palpable reminders of the canyon's history, surrounded, even while engaged in mundane tasks, by its remarkable beauty. It is not difficult to imagine this area inspiring the Navajo "First Song of Dawn Boy,"

> *Beauty before me,*
> *with it I wander.*
> *Beauty behind me,*
> *with it I wander.*
> *Beauty below me,*
> *with it I wander.*
> *Beauty above me,*
> *with it I wander.*
> *Beauty all around me,*
> *with it I wander.*
> *In old age traveling,*
> *with it I wander.*
> *On the beautiful trail I am,*
> *with it I wander.* [1]

The first human inhabitants here for whom there is archaeological evidence were migratory hunter-gatherers who entered the canyons of the de Chelly area nearly 5,000 years ago. Humans have inhabited the canyons at least seasonally ever since, producing the longest record of continuous habitation on the entire Colorado Plateau. The most visible artifacts of this long human history are spectacular cliff dwellings like White House Ruin, built by the Ancestral Puebloan people (known also as the Anasazi) in the abundant alcoves lining the canyon walls. These Puebloan cliff cities pulsated with activity between A.D. 750 and 1300, but they were abruptly abandoned for reasons that archaeologists still debate. Ancestral Pueblo descendants, the modern Hopi, migrated west to Black Mesa but still used these canyons for seasonal farming and religious purposes until the arrival of the Navajo, or Diné, as they refer to themselves, around A.D. 1700.

The Diné have lived here almost continuously since then, abandoning their beautiful and sacred canyons only briefly and under extreme duress in 1864, when Col. Kit Carson forced them on the Long Walk to New Mexico's Fort Sumner. After four years of incarceration, they were allowed to return to this area in 1868. Despite their past experience at the hands of the U.S. government, the tribe consented in 1931 to share

---

[1] As recounted in *Navajo Country*, by Donald Baars, University of New Mexico Press, 1995.

White House Ruin, one of many cliff dwellings in Canyon de Chelly

their home with all Americans as the country's first national monument jointly administered by the National Park Service and a Native American nation. The origin of the canyon's name, de Chelly, is lost in obscurity. Many ideas have been proposed, but the favored one at present is that it is an English corruption of a Spanish corruption of the Navajo name Tsegi (SAY-ih), which means, quite simply, "canyon." No matter what the name, Canyon de Chelly remains central to Diné spiritual identity.

As fascinating as the canyon's human history is, its stones have their own stories, too. Beautiful Canyon de Chelly was made possible by the rise of a large highland area known as the Defiance uplift, which forms the core of the Navajo homeland. The Defiance uplift is a broad dome that combines the Defiance Plateau, where Canyon de Chelly is located, with the Chuska mountain range to the northeast. Here on its western side, the dome slopes gently. It is bordered on the east by a steplike fold called the Defiance monocline (more about monoclines in vignette 5), which abruptly drops the area's colorful sedimentary layers into the valleys of western New Mexico. After the dome was uplifted, erosion carved Canyon de Chelly out of the top two sedimentary layers still present on the western plateau. In this vignette, you will climb from the surrounding flatlands up onto the Defiance Plateau, visiting several overlooks en route. At the last stop, hiking down to White House Ruin will give you a more intimate appreciation of the canyon.

The national monument visitor center is situated in the Chinle Valley near the toe of the Defiance Plateau. Heading to stop 1, you begin the noticeable climb up onto the plateau about 0.6 mile past the visitor center. As you pull into stop 1, you pass through a small roadcut composed of multicolored, crumbling rock almost devoid of vegetation. This outcrop is part of the Triassic-aged Chinle formation—the unit that also forms the pastel bedrock of the Petrified Forest National Park's Painted Desert and hosts the world's largest, most colorful deposits of petrified wood (vignette 15).

Below the mudstones that comprise the rim of the canyon at the overlook lies hard, pinkish, 234-million-year-old slickrock belonging to the Shinarump conglomerate, the layer that forms the surface of the Defiance Plateau throughout the national monument. This pebbly sandstone could not look more different from the crumbly mudstones above, but geologists group these two rocks together as the Chinle formation, with the Shinarump comprising the lowermost unit. The overlying mudstones that form

The great bulk of Black Mesa rising beyond this outcrop of typical Chinle formation (foreground) *at Tunnel Canyon Overlook (stop 1)*

Canyon de Chelly is carved out of only two rock layers. The dark, horizontally layered Shinarump conglomerate comprises the canyon rim and overlies the lighter-colored, prominently crossbedded DeChelly sandstone.

the bulk of the formation used to blanket the entire plateau, but as you can see, they are so easily eroded that they have been almost completely swept away.

Tunnel Canyon Overlook provides your first good look into Canyon de Chelly. The canyon here is diminutive, only about 200 feet deep, but its natural beauty is already apparent. The canyon is incised into the plateau's cap of Shinarump conglomerate as well as the top of the underlying DeChelly sandstone, a 275-million-year-old layer named for its breathtaking exposures here. Although both units usually look pink up close, you can distinguish the Shinarump cap by its occasional, well-rounded pebbles and, from a distance, by a slightly brown hue. In addition, the DeChelly sandstone contains prominent, sweeping crossbeds inclined up to 30 degrees, which the Shinarump lacks.

Higher on the western flank of the Defiance Plateau, Tsegi Overlook (stop 2) offers a panorama of the land to the west. Directly below you lies the valley of Chinle Wash, where the town of Chinle nestles. The wash has been carved into the mudstones you observed at stop 1. In the middle ground lie a series of low, red, flat-topped terraces. Beyond these rises a much taller mesa constructed of drab gray and tan rocks. This is Black Mesa, topping out at just over 8,000 feet, where some Navajo and Hopi families still make their homes (more on Black Mesa in vignette 5).

The roughly 5,000-foot stack of rock layers you see in Black Mesa used to continue up and over your head, over the top of the Defiance Plateau, and down the other side into New Mexico. Erosion stripped this impressive rock pile off the plateau, lowering the landscape down to the hard Shinarump conglomerate. If we could replace the entire stack, this area would top out at an elevation of roughly 11,000 feet, high above Black Mesa—yet another indication that you are climbing up the western flank of an ancient uplift as you travel eastward across the monument.

The Defiance Plateau rose to its prodigious height between 70 and 40 million years ago during a mountain-building episode known as the Laramide Orogeny. This event severely crumpled the earth's crust in Colorado and New Mexico, forming the Rocky Mountains. In the Four Corners

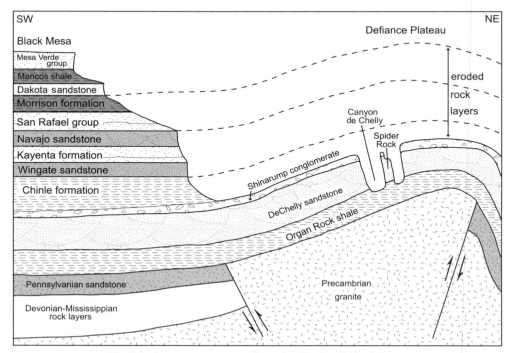

All of the layers now exposed on the flanks of Black Mesa used to arch up and over Canyon de Chelly.

area, however, it raised the Colorado Plateau straight up, preserving the horizontal nature of the area's sedimentary rock layers. The only exceptions to this even, elevator-like uplift were a series of broad domes, including the Defiance Plateau, which rose even further above the general elevation along monoclines. These single-sided folds formed when a few long-dormant faults in the underlying basement rocks rumbled back to life in response to the Laramide compression, buckling and uplifting the overlying sedimentary layers.

As you climb toward stop 3, the surrounding vegetation changes with the elevation. Tsegi Overlook's open land of scrub and sparse juniper trees gives way to a thicker piñon-juniper woodland; at Spider Rock Overlook, stop 3, a few tall ponderosa pine trees join the mix. As you walk the 200-yard path from the parking lot to the viewpoint, you will immediately notice that Canyon de Chelly has grown in stature, here forming a chasm 1,000 feet deep. In front of you, Spider Rock, one of the most impressive rock towers on the entire Colorado Plateau, rises sheerly from the junction of two canyons. This sacred spire is the home of Spider Woman, who taught the Diné how to weave and make clothes. You may well see signs of the canyon's current inhabitants in Spider Rock's shadow.

Spider Rock is sacred to the Diné, Canyon de Chelly's current inhabitants.

You should be able to distinguish Spider Rock's slightly darker cap of Shinarump conglomerate shielding the 825 feet of softer DeChelly sandstone below. At the spire's base you get a glimpse of a third layer, 280-million-year-old Organ Rock shale, which forms the pedestals of Monument Valley's famous towers (discussed in vignette 6). Geologists know from drill holes and a few outcrops farther east that this shale is the lowest layer of sedimentary rock on the Defiance Plateau. Beneath it lie Precambrian granite, schist, and quartzite basement rocks that geologists estimate to be 1,700 million years old. This enormous time gap of 1,400 million years is a surprise. In most other places on the Colorado Plateau, including the Grand Canyon, sedimentary layers the same age as these Canyon de Chelly rocks overlie at least another 3,000 feet of older sedimentary rock that accounts for at least 245 million years of this missing time.

From this discrepancy, we can guess that this is not the first time the Defiance Plateau has stood high above its surroundings. There must have been an older, ancestral Defiance Plateau. The lack of older sedimentary layers here can be explained in two ways. Perhaps the ancestral plateau was a topographically high feature for the entire 245 million years prior to deposition of the Organ Rock shale, causing these older sedimentary rocks to be deposited around but not over it. Or, those layers did in fact accumulate here, but were later eroded when, near the end of the time gap, the ancestral plateau was uplifted, causing erosion to accelerate. Given that a major period of mountain building swept the Four Corners region about 315 million years ago, the second explanation appears the most likely. At that time, a continental collision between northern South America and the Gulf of Mexico coast was raising a chain of mountains from Arkansas to New Mexico and Colorado. This collision was part of the assembly of the supercontinent Pangaea. The Colorado and New Mexico portion of this string of mountain ranges is known as the Ancestral Rockies, and the ancestral Defiance Plateau appears to have been a smaller outlier of that range. By 280 million years ago, erosion had planed the Defiance Plateau down to a small bump with Precambrian rocks forming its surface, allowing the layers of Organ Rock shale (coming from equally rapid erosion of the larger Ancestral Rockies to the east) and other sediments to be deposited over its top. These layers accumulated until 70 million years ago, when the modern plateau began its uplift.

On the hike down to White House Ruin at stop 4, we'll examine two of these layers up close. The trail is located in one of the best possible locations from which to piece together the history of deposition of the Shinarump conglomerate and DeChelly sandstone, the raw materials from which Canyon de Chelly is sculpted.

Twenty yards after you emerge from the tunnel that facilitates your descent off the rim, the trail takes a sharp, righthand switchback. Fifteen yards on from the switchback, next to a small tree to the right of the trail, are several large blocks of Shinarump conglomerate loaded with pebbles. These well-rounded pebbles offer you a clue to the environment in which this unit was deposited. Look back at the bedding at the tunnel's exit, a series of gently inclined layers that butt up against packages of horizontal beds. Known as low-angle crossbedding, this pattern is seen in sediment layers deposited in modern river and tidal channels, on portions of some beaches, and on shallow marine coastal shelves when storms stir up the sediments. Other clues along the trail will help narrow this list of possible watery origins.

Just to the left of the conglomerate blocks lie outcrops of soft, gray mudstone. Still considered part of the Shinarump, this softer rock has been more deeply eroded than the overlying conglomerate, creating an overhang of the harder rock above a shallow cave to the right of the trail. Clearly, the currents that laid down the Shinarump were at times strong enough to move pebbles, and at other times slack enough to let mud settle to the bottom. This wide range of sediment sizes rules out a beach environment, which is not prone to such variations in energy.

At its second switchback, the trail crosses onto the underlying DeChelly sandstone, which was deposited 275 million years ago. This rock looks similar to the less pebbly portions of the Shinarump along the canyon rim, but if you look closely, you'll see it is quite different. The DeChelly is devoid of pebbles and mudstone; it is well-sorted sandstone from top to bottom. Like the Shinarump, it is crossbedded, but these crossbeds are much larger and more steeply inclined. Such high-angle crossbeds form only in tall sand dunes, and there are just two environments where such dunes are found: deserts and coastal marine shelves swept by exceptionally strong tidal currents.

Just beyond the second switchback, the chiseled-out trail offers exceptional three-dimensional views of these features. Take a moment to examine some of the thousands of individual crossbeds, each thinner than a sheet of cardboard, comprising these steep walls. Notice how the crossbeds seem to come in groups separated by nearly horizontal bounding surfaces or thin stacks of horizontal layers. Geologists call each group of parallel crossbeds a cross-set. Many of the individual crossbeds—originally, each a small sand avalanche—run all the way down to the toe of the dune, where they merge out onto flat ground. You can see avalanche toes like these on the wall to the north, where most crossbeds curve down and left, merging with the bounding surfaces.

As sand transfers from the upwind to the downwind side, an entire dune shifts forward. Usually, a whole line of dunes is marching together in the same direction, one behind another. When a trailing dune moves forward and buries the flat area separating it from the preceding dune, this interdune area can be preserved as a thin package of horizontal layers between cross-sets, like the more vivid pink layers you see on the wall to the north. If these were tidal dunes, you would expect to find some marine fossils in these interdune deposits. However, no marine fossils have ever been found in the DeChelly sandstone. On the contrary, a large number of tracks made by beetles and rodentlike land animals were recently found in this layer near Monument Valley. Clearly, these dunes were deposited on land as desert dunes.

Such terrestrial dunes are always at the wind's mercy, and individual dunes last only as long as the wind keeps blowing in the same direction. Each cross-set represents the migration of a single dune before the wind recycled and reshaped the sand into a new one. Many dunes were

The cross-sets that identify the DeChelly sandstone as a sand dune deposit are clearly evident along the White House Ruin trail at stop 4. The bounding surfaces that separate cross-sets mark the boundaries between two separate dunes, and thicker packages of nearly horizontal layers are preserved interdune areas.

probably completely recycled, leaving no trace of their passage, but by counting the number of cross-sets here, you can arrive at a minimum number of dunes that migrated past this very spot. At least thirteen of them compose the cliff to the north.

Finally, note that virtually all of the crossbeds here tilt down to the left, or southwest. Since the crossbeds' orientations reveal which direction the wind was blowing when they were laid down 275 million years ago, we can tell that winds here blew steadily from the northeast.

Just 70 yards beyond the second switchback, you cross back onto the Shinarump conglomerate, darker brown here than it was above and loaded with pebbles and small cobbles. You can readily identify its contact with the pink DeChelly sandstone. Running through an alcove next to the trail, the contact is nearly vertical, but just above the alcove it is a nearly horizontal, gently undulating surface. What could account for this unusual rock juxtaposition?

The only plausible explanation is that you are standing in an ancient channel. There are several possible ways such a channel could form. It could have been carved by a river, by oceanic tides, or by hurricane-force storms ravaging a shallow marine shelf. As illustrated by the contact, the channel had nearly vertical walls, and it was almost 100 feet deep, stretching from the present canyon rim to below the next switchback. If you examine the slope immediately below the trail, you will find fist-sized chunks of pink DeChelly sandstone embedded in the conglomerate. These chunks of DeChelly sandstone were ripped off the channel walls by a swift current that roared down this very deep channel. Such deep channels and swift currents don't fit with a storm deposit on a marine shelf, so we have narrowed our list of depositional choices to just two: a river or a tidal channel. As in the DeChelly, no marine fossils have ever been found in the Shinarump. Further, ripple marks and other indicators of ancient flow direction reveal that the waters in this channel, as well as others in the Shinarump, ran only in one direction, not two, as you would expect in a tidal channel. These clues, among others, have led geologists to conclude that a series of swiftly flowing streams deposited the Shinarump.

By now we know that the Canyon de Chelly area was uplifted during the continental collision that created the supercontinent Pangaea; that by 280 million years ago, it had been reduced to a broad, flat river floodplain; and that 275 million years ago, a drying of the climate ushered in sand dunes that blew across the landscape. By 234 million years ago, the climate had changed again, and swift, steep rivers coursed across the area. More sedimentary layers accumulated until 70 million years ago, when the modern Defiance Plateau uplifted and its upper layers—a stack of sedimentary rocks 5,000 feet thick—began to erode.

A channel of dark, pebbly Shinarump conglomerate cuts deeply into the lighter DeChelly sandstone layers along the White House Ruin trail. This subsidiary channel lies adjacent to a much deeper channel that crosses the trail.

Blocks of DeChelly sandstone in the Shinarump conglomerate

But one mystery remains. What happened during the time gap between deposition of the 275-million-year-old DeChelly sandstone and the 234-million-year-old Shinarump? In most places—for example, in Monument Valley, as we saw in vignette 6—one or more rock layers lie between the Shinarump and the DeChelly. Where are those intervening layers here on the Defiance Plateau? Were they never deposited? Or were they laid down and then removed before the Shinarump streams flowed through?

During the 41 million years between deposition of the DeChelly and the carving of this canyon into it, percolating groundwater had at least partially cemented the sandstone into rock. We know this because the DeChelly supported vertical channel walls and because the chunks of sandstone that broke off during floods did not completely disintegrate as they tumbled around in the river. Such cementation usually occurs after a sediment body has been buried beneath a significant thickness of overlying sediment. At least some sediment layers lay above the DeChelly before the Shinarump-depositing rivers came roaring through.

The trail switches back to the right before intersecting this ancient channel one final time. The rest of the way to the bottom of the canyon, you pass nothing but DeChelly sandstone, with marvelous views of its stunning cross-sets. The trail crosses the wash and ends at White House Ruin, the beautiful Puebloan cliff dwelling sacred to the Navajo. The ruin is set in an alcove chipped out of the DeChelly sandstone along a line of springs. The springs exit the wall in a former interdune area composed of silt that, due to the overlapping, shingling effect of its platelike mineral grains, effectively stops further downward percolation of the groundwater. As each drop of water flows out of the ground, it carries a few sand particles with it. This process, called spring sapping (more in vignette 5), is slow, but it can eventually carve out an alcove as large as this one.

Canyon de Chelly is not the biggest, longest, or deepest canyon in Arizona, but it is unquestionably one of the most beautiful. The simple elegance of its architecture, owing to the fact that it is carved from only two rock layers, enhances its appeal. To keen-eyed visitors, these two layers offer abundant clues to their environments of deposition and hence the ancient geographies that existed here long ago. Recognition of this deep geologic history adds another dimension to the canyon's rich human history. Both are integral to the special ambiance of Canyon de Chelly; both are there to absorb as you "wander with beauty all around."

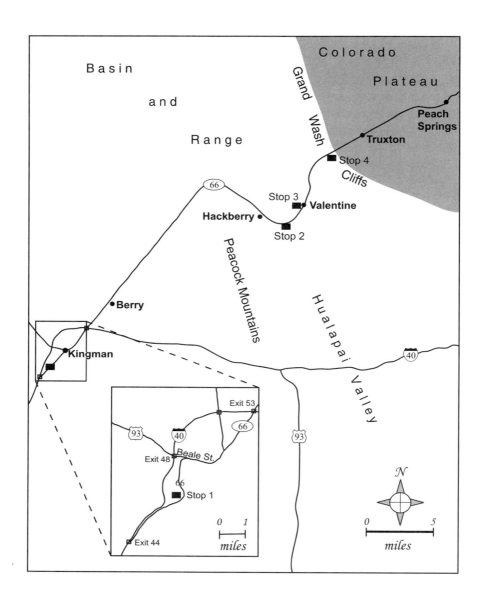

✦ GETTING THERE

This vignette features four stops between Kingman and Peach Springs in the Hualapai Valley in northwestern Arizona—a total distance of about 40 miles. To get to stop 1 from I-40, take exit 48 (Beale Street) in Kingman and turn right (southeast). After 0.3 mile on Beale Street, turn right (south) onto historic Route 66, also called the Oatman Highway, which immediately enters a small gorge carved into the Peach Springs tuff. Stop 1 is at milepost 48, 1.5 miles down the road; there are small pullouts on either side of the road. To reach stop 2, get back on I-40 heading east. Take exit 53 (Andy Devine Avenue—yet another name for Route 66), turn left (north), and follow Route 66 for 24 miles to Hackberry. About 2 miles beyond the town and 0.5 mile after milepost 82, pass the junction with Hackberry

10
Volcanic Violence in a Landscape Turned Upside Down
THE PEACH SPRINGS TUFF

About 18.5 million years ago, fire and brimstone from an almost unimaginable volcanic cataclysm engulfed the Kingman area. Today, remnants of the rock that erupted during that violent episode are still scattered across an astounding 13,000 square miles—an area nearly as large as Massachusetts, Connecticut, and Rhode Island combined. This rock, known as the Peach Springs tuff, is found in California, Nevada, and Arizona. It varies in thickness from just a few feet at its edges to well over 300 feet around Kingman and in eastern California. Based on these details, geologists estimate that the eruption was much larger than any explosion in recorded human history; it expelled about 150 cubic miles of volcanic ash—a staggering 625 times more material than Mount Saint Helens ejected in 1980. The Peach Springs tuff has a lot to tell us about what this area was like at that time and what happened later.

After parking at stop 1, examine the fresh roadcut on the right side of the highway and the big, quarried blocks on the left. All the rock you see here is part of the Peach Springs tuff. The tuff's great thickness is a vivid reminder of how much material was expelled during the brief but powerful outburst. The massive event was known as a caldera eruption. During such an eruption, huge pressure empties the entire magma chamber within a volcanic mountain. Since this empty space can no longer support the mountain, it collapses, leaving a huge circular depression—a caldera.

Caldera eruptions are the most violent and explosive type of volcanic activity. They spew enormous quantities of atomized lava into the air,

---

Road (Mojave County 141) and park in a small pullout immediately on the right. Look for the low roadcut across the road. For stop 3, continue east 2.5 miles on Route 66 to Valentine. At milepost 85, turn left (west) and follow the short access road to the Bureau of Indian Affairs office. Park at the bank of mailboxes in front of the office. Stop 4 is 6 miles further east along Route 66. Just beyond milepost 91, a series of small, undulating hills reveals three small roadcuts in quick succession on both sides of the road. Park on the right in a pullout beyond the third roadcut, 0.4 mile past milepost 91.

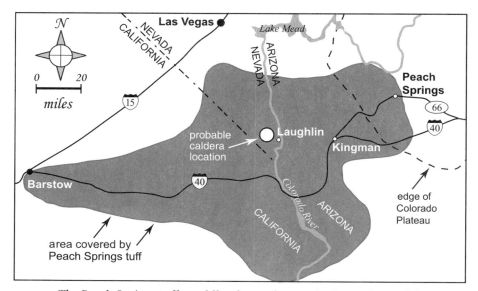

The Peach Springs tuff straddles the modern geologic provinces of the Colorado Plateau (north and east of the plateau's edge) and Basin and Range (south and west of the edge).

where it solidifies into scorching particles of ash. Smaller, lighter particles are injected miles into the atmosphere, where strong high-altitude winds catch them up and blow them around the world.

Much of the ash, though, falls to the ground like a gentle snowfall in the hours to weeks following an eruption, blanketing the landscape with anywhere from several feet near the volcano down to a dusting hundreds of miles away. These deposits, known as ash falls, usually form a crumbly layer of volcanic rock with small, very uniform particles.

The particles in this roadcut, however, are not all the same size, nor is the rock loosely consolidated, or crumbly. A light-gray to pink background, or matrix, surrounds two different sets of larger embedded particles, similar to raisins and nuts baked into a loaf of bread. These particles are too large and this deposit is too hard and thick to be from an ash fall.

Instead, an ash flow left these rocks. When calderas erupt, clouds of larger ash particles too heavy to be blown high into the sky fall back to earth, trapping a layer of superheated air beneath them. This thick mass of scalding material rolls down the volcano's flanks at speeds greater than 60 miles per hour, gliding on a cushion of air like an air-hockey puck on its game table. Ash flows hug the topographic lows, just like a river, surging down valleys and other depressions, mercilessly destroying everything in their path.

As the flow slows and stops, 1,200-degree-Fahrenheit ash at the bottom is welded together by the fierce heat and crushing pressure of the overlying pile. This strong, resistant rock is called welded tuff. In the field, ash flow deposits like this one can be distinguished from ash fall deposits by their greater thickness, less crumbly nature, and larger, more variably sized particles.

Look closely at the two varieties of larger embedded particles. The first type consists of rock fragments: jagged, light brown chunks of pumice, most about 0.5 inch across. Pumice forms when magma full of trapped gases is quickly ejected from a volcano. As the lava cools, the trapped gases escape into the atmosphere, leaving behind a glassy volcanic rock so full of enclosed bubbles that the rock, less dense than water, can usually float.

The fact that these pumice chunks are embedded in the matrix of tuff narrates a tale of volcanic violence involving not just one, but multiple eruptions. The pumice chunks were first created in an earlier outburst and deposited in a layer that formed the top of the volcano. When the

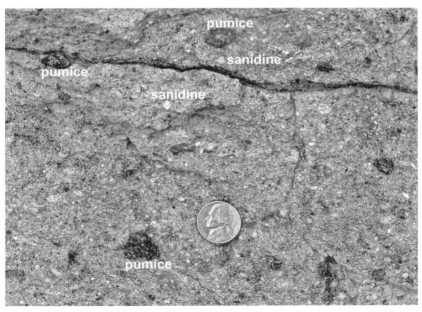

The Peach Springs tuff at stop 1. Note the dark, embedded chunks of pumice (a type of volcanic glass) and the lighter, rectangular crystals of the mineral sanidine in the more uniform matrix of tuff.

"big one" occurred 18.5 million years ago, the eruption was so powerful that it blasted some of these earlier rocks into the air along with the molten material. The pumice fragments tumbled along with the erupted ash, where they remain embedded to this day.

The second group of embedded particles, or inclusions, consists of small, rectangular, white crystals that sometimes flash pale blue. These are crystals of the mineral sanidine. Their relatively large size and angular shape indicate that they grew very slowly in the magma chamber prior to the eruption. During the explosion, they and the pumice were ejected along with the magma, which cooled very quickly to form the pink, glassy matrix now entombing both types of inclusions.

The style of any volcanic activity is strongly influenced by how much silica the erupting magma contains, as well as how much gas rises through the earth's crust along with the magma. Silica is a compound of silicon and oxygen that comprises 45 to 75 percent of most crustal rocks.

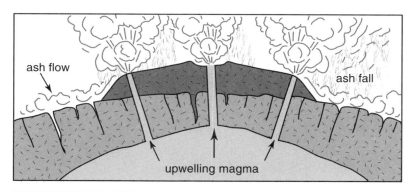

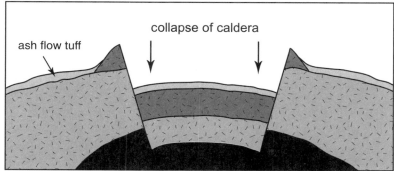

Caldera eruptions empty the entire magma chamber feeding the volcano; the weight of the overlying material causes it to collapse into the newly created space.

Sanidine only crystallizes from magmas with a very high proportion of silica, so these sparkly crystals, as well as the tuff's light color, tell us that the source material contained a very high proportion—more than 70 percent.

The gases in a magma chamber behave like carbonation in a can of soda. Low-silica magma is quite runny, so the rising gases can easily escape, the way they do from an open can of soda that rapidly goes flat. However, high-silica magma is especially sticky and thick. It traps the gas bubbles, which build up over time, increasing the pressure in the magma. The effect is similar to vigorously shaking a sealed can of carbonated soda. Pop the top, and the contents, under high gas pressure, spew out: a violent, sticky soda eruption! Under a high-silica volcano such as the one that existed near here, the same type of process occurs. Gas pressure builds up in the magma chamber until it exceeds the strength of the overlying rocks. They fracture and release in a tremendous explosion. If the eruption releases a significant portion of the magma under the volcano, we have a caldera eruption.

Because fault activity and erosion since the eruption have obscured the caldera that produced the Peach Springs tuff, its exact location and size are not known with certainty. But the tuff's thickness decreases both to the east and west of the Colorado River corridor, and geologists have long speculated that the Peach Springs caldera lies in southernmost Nevada near today's city of Laughlin. Recent magnetic studies corroborate this speculation. The tuff contains the magnetic mineral magnetite, the crystals of which were aligned along the flow direction of the ash. Ash flows raced away from the caldera radially in all directions. By measuring the magnetic orientation of these magnetite crystals, scientists traced the flow lines back to an area that may be the location of the source caldera.

Based on the huge volume of rock it produced, the Peach Springs caldera probably measured 10 to 12 miles across—comparable to Tanzania's famous Ngorongoro Crater, currently the sixth largest in the world.

The explosion of the Peach Springs caldera was merely one event, albeit one of the largest, in a massive wave of volcanic activity that "rocked" the southwestern United States beginning about 30 million years ago. Geologists attribute this pulse of volcanism to a major reorganization of tectonic plates off the coast of California at that time, which had profound impacts on the geography and geology of the entire region. It leveled a mighty mountain range. It gave birth to California's notorious San Andreas fault. And it opened up the Basin and Range province, the landscape we see here and throughout western Arizona and adjoining portions of California, Nevada, and Utah, with its countless steep, north-south-trending mountain ranges and intervening dry,

Tanzania's Ngorongoro Crater, about the same size as the Peach Springs caldera

The Peach Springs tuff in a spectacular series of roadcuts along I-40. Banding of the tuff here results from slight differences in the chemistry of the magma ejected during different stages of the eruption. Small normal faults have offset these bands.

pancake-flat valleys. More than any other single event, geologists have found, this plate reorganization is responsible for sculpting the modern face of the American Southwest.

Because the Peach Springs tuff instantly blanketed much of the area affected by this tectonic event, it serves as a crucial "marker bed" for tracking how the landscape changed in response to plate reorganization. To unravel any rising, falling, or tilting the region has endured since the great eruption, we can carefully observe the tuff, its relationship to surrounding rocks, and the elevation of each outcrop. As we do some of this tracking, just ahead, bring along an altimeter (an instrument that

measures vertical distance from a reference point) or a Global Positioning System device if you like. Altimeters should be calibrated in Kingman, which sits at an elevation of 3,340 feet in the Hualapai Valley, the easternmost valley in Arizona's portion of the Basin and Range.

At stop 2, take a moment to orient yourself. West of you lie the Peacock Mountains, the easternmost chain in the Basin and Range in this area. To the east lies a bold escarpment known as the Grand Wash Cliffs, the western edge of the Colorado Plateau. You are standing in the narrow transition zone between these two spectacular geologic provinces.

The near (west) end of the roadcut across the highway consists of layers of loosely consolidated river gravels loaded with cobbles of black basalt, a fine-grained volcanic rock. In these gravels you will also see many glassy, brown fragments of our old friend, the Peach Springs tuff, demonstrating that these gravels were deposited *after* the tuff was deposited and began to erode. The gravels must therefore be younger than 18.5 million years. Most sedimentary units like this are deposited very close to horizontally, and in a tectonically quiet area they will remain essentially flat. However, in a tectonically active area, they can be tilted and even overturned. These layers are clearly tilted down toward the right (east), indicating that some sort of tectonic movement occurred here since the Peach Springs eruption.

A short stroll to the outcrop's eastern end reveals the cause of this tilting. The gravels' prominent layers end abruptly against a bright red, steeply tilted zone of highly weathered rock. This is a normal fault—a break in the rock that makes room for extension, or stretching, of the earth's crust. On the fault's right side lies granite. This rock's hallmark interlocking white feldspar and clear quartz crystals are easier to recognize if you walk another 30 feet east.

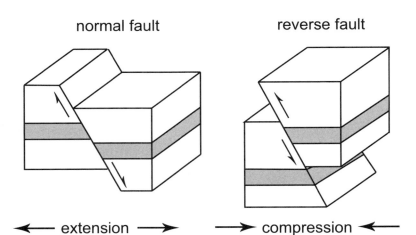

Two types of faults. Normal faults, in which the upper block of rock slides down relative to the lower block, are associated with tectonic extension, which formed the Basin and Range. Reverse faults, in which the upper block of rock slides up and over the lower block, are associated with tectonic compression, which uplifted a large mountain range where the Mojave Desert is now.

A family of faults similar to this one accomplished huge amounts of stretching during the opening of the Basin and Range, and similar outcrops allowed geologists to piece together the saga of tectonic reorganization. Off the coast of California, a tectonic plate boundary known as a subduction zone existed for 200 million years. There, two tectonic plates met; one of them, colder and thus denser, dove beneath the other (the subduction process is described in vignette 16). Once the descending plate plunged deep enough to melt, resulting magma blobs began rising to the surface through the earth, feeding a chain of high volcanic mountains where the Sierra Nevada range now lies.

Because a descending plate is mercilessly squeezed and doesn't yield gracefully, subduction zones are tectonically fierce areas. Subduction off the coast of California also thrust up a tangled chain of very high, nonvolcanic mountains, known as the Sevier-Mogollon Highlands, east of the ancestral Sierras in the region now occupied by the Mojave Desert and the Basin and Range province. Based on reconstructions of the thickness of their crust, it is believed that, in their prime, they were as high as the modern Andes, with summits likely soaring over 20,000 feet.

As the subducting plate, known as the Farallon plate, slowly disappeared, a mid-ocean ridge on its trailing western edge was dragged ever closer to California. As a buoyant area where new oceanic crust is forged,

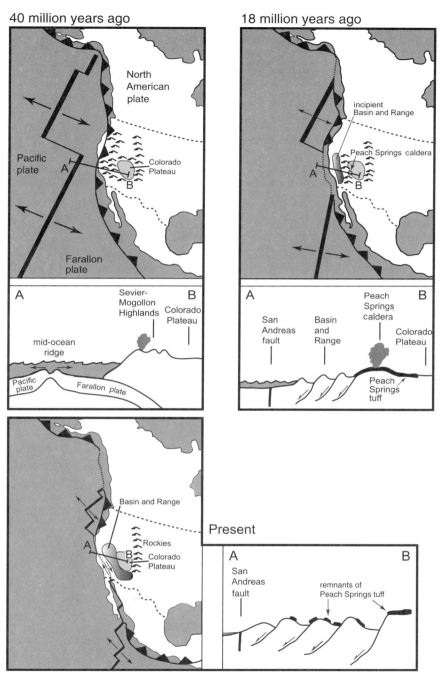

The topographic evolution of the American Southwest. Each map is paired with a cross-section from point A to point B. Arrows show direction of fault movement. About 40 million years ago, a subduction zone (barbed line) ran along the entire west coast of North America, uplifting the Sevier-Mogollon Highlands in western Arizona, and a mid-ocean spreading ridge (thick line) drew ever closer. By 18 million years ago, the spreading ridge had met the subduction zone in southern California, giving birth to the San Andreas fault (finely dotted line). This tectonic reorganization triggered growth of the Basin and Range province through collapse of the Sevier-Mogollon Highlands and fueled volcanic eruptions, including those from the Peach Springs caldera. Continued opening of the Basin and Range led to the present topography.

this ridge separated the downgoing Farallon from its western neighbor, the Pacific plate. Finally, approximately 30 million years ago, the ridge entered the subduction zone off the coast of Los Angeles. For the first time ever, the Pacific plate came into direct contact with the North American plate, and the tectonic activity switched from subduction to sideways sliding between the new neighbors along an important new fault: the San Andreas.

Since formation of the San Andreas fault, the Peach Springs tuff tells us, the most dramatic change has been the birth and expansion of the Basin and Range where the towering Sevier-Mogollon Highlands once stood. The roadcut you are examining is one of the actual faults that chopped up the ancient highlands. Forced up by the great compression occurring to the west, the highlands consisted of massive amounts of heavy rock thrust high in the air. Gravity constantly tugs on such large mountain ranges, trying to tear them down. As long as subduction continued, the squeezing combated the force of gravity. When the subduction zone died, however, gravity gained the upper hand, and the mountains collapsed from their own weight, foundering along a series of north-south-trending faults that sliced through them. As one side of each fault slid down, a series of valleys opened up, chopping the highlands into smaller ranges separated by parallel valleys, and significantly lowering the region's average elevation.

Movement along the normal faults stretched and thinned the earth's crust like taffy. This allowed the earth's hot mantle to rise unusually close to the surface. That triggered large-scale melting of rock and the formation of numerous magma chambers, which in turn spawned hundreds of volcanoes, enveloping the area with lava and ash. One of these chambers was the source of the massive Peach Springs eruption. This tuff is a direct result of the death of a subduction zone off the west coast 30 million years ago.

Movement along normal faults like this one has dramatically tilted many of the rocks in the Basin and Range. This is apparent in both the gravels in front of you and a series of tilted layers of basalt and Peach Springs tuff visible to the south. In contrast, farther east you can see the flat-lying layers of the Colorado Plateau. At this outcrop you can actually put your finger on the boundary between these two great provinces.

The fact that movement along this normal fault has tilted layers of Peach Springs Tuff and cut even younger gravel deposits demonstrates that this portion of the Basin and Range formed more recently than 18.5 million years ago. But what did the landscape here look like before normal faulting formed the Hualapai Valley? Part of the answer lies ahead at Valentine, stop 3.

Traveling east from the fault at stop 2, Route 66 climbs onto the Colorado Plateau. Most of the rocks around you are the same granites you saw near the fault. However, on the tops of the mesas, you can see a bold, blocky cliff band that is obviously bedded and nearly horizontal. It's a 100-foot-thick layer of Peach Springs tuff—much thinner than it was in Kingman.

At stop 3, look toward the mesa east of the highway. The same prominent horizontal band of Peach Springs tuff you saw on the way here caps the mesa. On the mesa's right side, directly below the tuff, are prominent outcrops of hard, nearly white rock. This is the same ancient granite you've been driving through. Similar granite outcrops underlie the tuff on the mesa's left side. Below the distinctive tuff band in the center, however, you see not granite, but rather a series of dark, horizontally banded rocks peeking out from beneath sparse vegetation. These are layers of volcanic and sedimentary rocks. To both the left and the right, the horizontal

At stop 3, horizontal layers capped by the cliff of Peach Springs tuff fill an ancient river channel carved into the lighter-colored and much older granite below. At the time of the Peach Springs eruption, this channel drained the Sevier-Mogollon Highlands to the west.

bands end abruptly against outcrops of white granite that seem to angle downward toward the center of the mesa, like a giant letter U.

The U shape outlines a paleocanyon—an ancient canyon—that erosion carved into the granite long ago. Over a period lasting several million years, this paleocanyon filled with flat-lying sediments and lava flows from other volcanic activity, gradually reducing it to a low swale. During the massive Peach Springs eruption, ash filled and even overtopped this canyon. Geologists scouring the area have found many similar paleocanyons filled with the same outcrops. In deeper canyons, additional layers of younger volcanic and sedimentary rocks bury the tuff.

So at Valentine, the Peach Springs tuff presents us with an important clue regarding the lay of the land 18.5 million years ago. The Colorado Plateau was cut by a series of deep canyons that, over millions of years, filled with volcanic and sedimentary material. Dating of volcanic rocks in the fill indicates that these paleocanyons were carved and began to fill with sediment around 65 million years ago, about the same time the Colorado Plateau began to rise. The deepest ones weren't completely buried until 5 to 6 million years ago, finally leveling the area to form the relatively flat landscape we see today.

Notice that the tuff cliff looming above you is about 100 feet tall, only a third of the thickness of the tuff in Kingman. If the tuff here consists of ash flow deposits as it did at Kingman, that suggests Valentine lay farther from the caldera, and downhill, at the time of the eruption. Yet the base of the Peach Springs tuff above you sits at about 4,500 feet, considerably higher than in Kingman. If this is indeed an ash flow deposit, why is it higher here than at Kingman, which is closer to its source? Unfortunately, we can't access the tuff here to verify which type of deposit it is, but we'll be able to answer the question at stop 4.

As you again head east on Route 66 from Valentine, the road draws closer to the tuff. From the parking area at stop 4, walk back along the road 100 yards to the series of three small roadcuts. All of the layers in these cuts are tilted gently down to the east, so each layer you encounter as you walk west is older than the previous one.

In the middle cut, the familiar Peach Springs tuff lies sandwiched between gravels. Quick reconnaissance uncovers the same rectangular, white sanidine crystals and large chunks of pumice seen at stop 1. Their presence demonstrates that this is indeed an ash flow deposit identical to the one you examined near Kingman, with one exception: here, near the town of Peach Springs, the tuff is just a scant few feet thick—a dramatic thinning from the 100-foot cliff at Valentine just 6 miles southwest.

The fact that the ash flow was 300 feet thick in Kingman, but is only a few feet thick here, indicates that this location lay near the tail end of the flow. Because gravity drives their movement, ash flows surge downhill, so this spot must have lain a good deal lower than Kingman at the time of the eruption. But today this valley sits at 4,250 feet, over 900 feet higher than Kingman. Since the tuff was erupted, what used to be higher in elevation is now lower, and vice versa.

By giving us a way to judge elevation changes, the Peach Springs tuff allows us to paint a compelling portrait of this ancient landscape. The erupting Peach Springs caldera lay high in the mountains of the Sevier-Mogollon Highlands, in present-day Nevada. Ash belched out by the eruption flowed east down the flanks of the volcano and surged through a series of canyons, like the one you saw at Valentine, carved into the range's low foothills. The ever-thinning flow of ash finally halted a few miles north and east of here.

Geologists who have traced the tuff west into California have completed this landscape sketch for us. In the Mojave Desert, outcrops of Peach Springs tuff lie nearly horizontal over the top of steeply tilted volcanic rocks about 2 million years older. This relationship indicates that in southern California, the Sevier-Mogollon Highlands had already collapsed to form the tilted mountain ranges of the Basin and Range *before* the eruption 18.5 million years ago.

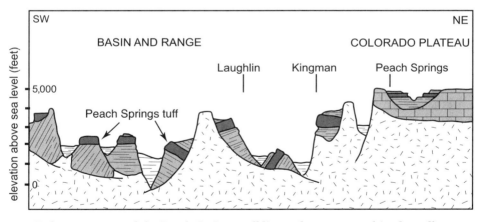

Today, remnants of the Peach Springs tuff lie on the ranges and in the valleys of the Basin and Range province and fill ancient river channels carved into the edge of the Colorado Plateau. In California (left side of figure), the tuff rests horizontally above strongly tilted rocks, indicating that the western Basin and Range formed before 18.5 million years ago. In contrast, the tuff in Arizona tilts with the rocks below it, documenting the collapse 18.5 million years ago or less of the Sevier-Mogollon Highlands and formation of the eastern Basin and Range.

In contrast, many of the Peach Springs tuff outcrops in Arizona, like those south of Hackberry (stop 2), are strongly tilted. This shows that in Arizona, the collapse of the highlands didn't occur until after the big eruption. At that time, western Arizona stood considerably higher than northern Arizona. In the geologically short interval since the Peach Springs eruption, the topography between northern and western Arizona has completely reversed. Now the north stands tall and the west has collapsed into a low, barren desert.

The phrase "as old as the hills" is a reminder that mountains are a strong symbol of longevity for human beings. Mountains do indeed far outlast our fragile human works, but as the Peach Springs tuff shows, even mountains fall. Since the Peach Springs tuff was erupted, only four-thousandths of the time since the earth was born has elapsed. But in that geologically brief span, as we have seen in our travels along the path of the tuff's flow, western Arizona was transformed from a landscape of lofty peaks to the hot, dry valleys of today's Mojave Desert.

11
A Mélange of Magmas, a Variety of Volcanoes
THE SAN FRANCISCO VOLCANIC FIELD

Flagstaff is the hub of Arizona's high country, a mountainous region with peaks so lofty that one even supports a ski area—an anomaly in a state noted more for its saguaro cacti than for skiing. Over six hundred individual hills and mountains rise above the high plateau on which Flagstaff sits. From conical 100-foot hills to steep, majestic Humphreys Peak—Arizona's highest peak at 12,633 feet—every one of these landforms is a volcano, part of the 1,800-square-mile San Francisco volcanic field. In few places in the nation can you observe so many different types of volcanoes in such a small area. This exceptional variety of volcanoes can be traced to the mélange of magmas (molten rock inside the earth) that fed them.

Sunset Crater National Monument is a great place to start your exploration of the area's volcanic diversity. Volcanism in the San Francisco volcanic field began roughly 6 million years ago, and volcanoes have been erupting sporadically ever since. At 950 years of age, Sunset Crater

Volcanoes in the eastern portion of the San Francisco volcanic field

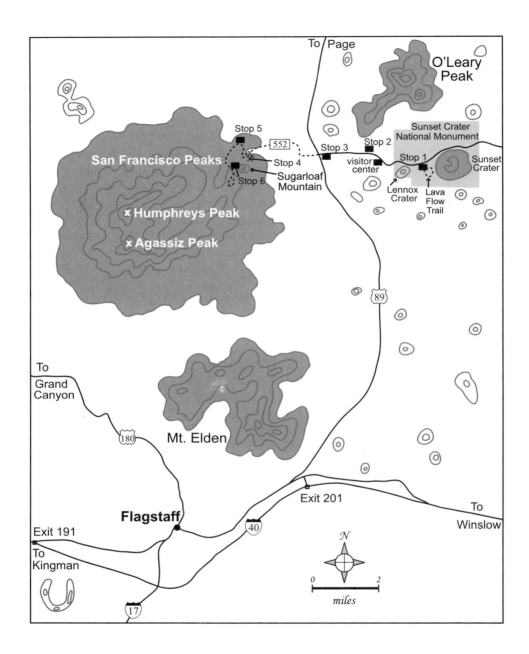

✦ GETTING THERE

This vignette takes you to six stops in the eastern portion of the San Francisco volcanic field north of Flagstaff. Stop 1 is the Lava Flow trailhead in Sunset Crater National Monument. Drive 12 miles north of Flagstaff on U.S. 89 to the sign for Sunset Crater and Wupatki National Monuments. Turn right; drive for 2 miles to the entrance station and visitor center, then continue another 1.5 miles east to the Lava Flow Trail parking area. You can tour the

Close-up, stops 3–6

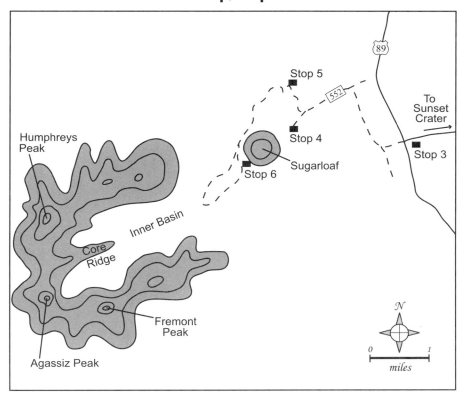

lava flow here on either of two loop hikes, 0.25 mile or 1 mile long, that share the same trailhead. For stop 2, from the Lava Flow parking area, backtrack west 2.4 miles to one of the scenic pullouts on either side of the road in Bonito Meadow. Continuing west, stop 3 is a pullout on the left (south) side of the road just before it joins U.S. 89. To reach stop 4, an abandoned pumice quarry on the flanks of Sugarloaf Mountain, continue west, crossing U.S. 89 onto Forest Road 552, a graded dirt road that is clearly signed to Lockett Meadow. Turn right at the T-junction 0.5 mile from U.S. 89, and left at the next T. After 1.6 miles from the highway, FR 552 turns right at a sign to Lockett Meadow; continue straight here on a spur off FR 552 for 0.2 mile and pass through an open gate to the quarry. To reach stop 5, go back to FR 552 and turn left. Immediately pass through a gate, if it's open (it may be closed from about November through April). This steep mountain road is in good shape but quite narrow in spots. Stop 5 is a wide scenic pullout on the right at a sweeping left bend 0.9 mile from the gate. After 2 additional miles, you reach the beginning of a one-way loop in Lockett Meadow. Stop 6 is 0.7 mile around the loop at campsite 17.

Sunset Crater is a classic cinder cone. The black, open area in front of the cone is the Bonito lava flow (stop 1); the lighter clearing is Bonito Meadow (stop 2).

is the youngest volcano in the field. From a human perspective, that seems quite ancient, but by volcanic standards, Sunset Crater is a spring chicken. The presence of such a young volcano strongly suggests there is still magma below the surface here. At a seismic monitoring station in neighboring Wupatki National Monument, scientists confirmed this hypothesis by observing earthquake-generated sound waves slow down as they passed through magma chambers beneath the volcanic field. Sunset Crater is unlikely to be the last volcano to erupt in these parts.

On the way to stop 1, breaks in the ponderosa pine forest reveal tantalizing glimpses of this beautifully symmetric crater. Near the Lava Flow trailhead, however, the stately pines suddenly disappear. In their stead is the startlingly black and barren Bonito lava flow, so fresh looking it appears to have erupted a scant few years ago.

When you step out onto the lava at the Lava Flow trailhead (stop 1), your first impression may well be one of harsh, dark, and jumbled terrain that feels utterly foreign—more like the moon than the earth. In fact, Apollo astronauts trained for their missions to the moon in the San Francisco volcanic field and nearby Meteor Crater (vignette 14).

From the parking lot, follow the trail toward Sunset Crater until you reach a bridge over a shallow crack that split open as the fresh lava chilled. Like all volcanoes, Sunset Crater erupted when magma wormed its way up through the earth's crust and spewed onto the surface (magma's name changes to lava after it reaches the surface). When it erupts, lava is incredibly hot, but freed of earth's fiery furnace it cools quickly and unevenly, forming a crumpled crust and large cracks like this one.

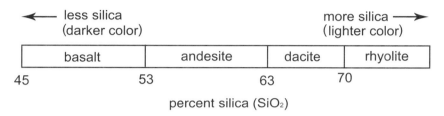

The classification system for volcanic rocks based on their silica composition

The Bonito lava flow is composed of jet-black rock classified by geologists as basalt. Volcanic rocks are some of the easiest to identify in the field because they're classified by color, which magma's chemical makeup ultimately controls. The main ingredient in most magma recipes is a pyramid-shaped compound called silica, with one silicon atom sandwiched between four oxygen atoms. Other rock-forming elements such as iron and magnesium encircle these fundamental silica building blocks to form a tacky dough. As with most recipes, the proportions of the ingredients have a lot to do with what the end product is like. Add too little baking soda in a cake, and it will be as flat as a board. For a volcano, the most important ingredient is silica, the proportion of which can vary from around 45 percent to over 70 percent for rocks in earth's crust. The more silica a volcanic rock contains, the lighter its color. The blackness of the rock here indicates very little silica is present.

Knowing magma's silica content is crucial because it, along with the temperature of eruption, controls the lava's resistance to flow—its viscosity. Viscosity is the single most important factor determining a volcano's style of eruption, and consequently its size and shape. Basaltic lava, with its low silica content (45 to 53 percent), is quite runny as lava goes; more viscous than molasses, it can nevertheless flow for miles and succeeds in releasing most of the gases pent up in it along the way. The greater the buildup of these gases, the bigger the volcanic explosion, so basalt eruptions, with their modest buildup of trapped gases, are relatively tame—at least compared to eruptions like Mount Saint Helens in 1980.

Still, farmers of the prehistoric Sinagua culture who witnessed Sunset Crater's birth probably thought that eruption was exciting enough. Tree-ring experts have determined both from the beams of ancient Sinaguan pit houses buried by lava flows and from trees living at the time that the fireworks began in late A.D. 1064, or early 1065 before the growing season. The eruption started with glowing red lava fountains spurting from a 6- to 9-mile-long crack, a type of outpouring that scientists describe vividly as a "curtain of fire." Due to the pressure of the overlying material, gases at depth are dissolved within magma. However, when the magma wells up to the earth's surface during an eruption, the pressure is reduced and the gases expand as bubbles, as they do when you crack open a can of soda. If there are enough gases, their expansion can fling magma up to 2,000 feet into the air.

As this initial phase waned, the eruption narrowed to one end of the crack, where a 1,000-foot mountain grew. The lava sprays cooled rapidly as they fell to earth, solidifying into rough fragments called cinders. Just as popcorn collects around an old-fashioned hot-air popper if you remove its lid, the heaviest cinders piled up around the central vent to form Sunset Crater.

As the eruptions progressed intermittently for over one hundred years, the magma's gas content decreased until it could no longer sustain the lava fountains. In A.D. 1180, lava instead seeped from an opening on the cone's northwest side into a small basin, where it pooled and hardened to form the Bonito lava flow. In a final burst of activity around the year 1200, lava containing abundant iron and sulfur erupted. It quickly rusted in the atmosphere and fell back on the crater, dusting the volcano's top with the red and yellow "sunset" hues for which it is named.

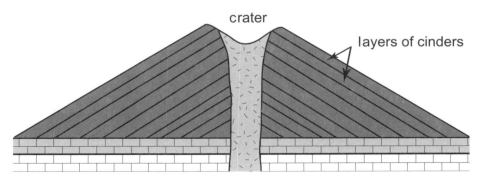

A cinder cone like Sunset Crater is formed of layer upon layer of cinders ejected from the central crater.

Basalt of the Bonito lava flow is filled with vesicles from trapped gas bubbles.

Look for evidence of once active subterranean gases around you on the Lava Flow Trail. Loose cinders are scattered over the top of the lava flow. Peppering many of the rocks are small, rounded holes known as vesicles. These are the remnants of trapped gas bubbles "frozen" in the rock as the lava solidified.

On your way out of Sunset Crater National Monument, the visitor center is well worth a trip, with exhibits on the Sunset Crater eruption, the variety of volcanoes in the San Francisco field, and the earthquakes detected at Wupatki. One particularly evocative artifact on display is an impression left by a Sinaguan corncob in basalt of the Bonito flow.

Most of the San Francisco field volcanoes are cinder cones like Sunset Crater, but there are several big (quite literally) exceptions. Foremost among these is dramatic San Francisco Mountain, the remnant of a much larger and more explosive volcano. A string of half a dozen intermediate-sized volcanoes also stretches across the area. All of the taller mountains owe their lofty heights to the fact that, unlike the cinder cones, they did not erupt basalt. Instead, the intermediate-sized volcanoes are composed of lavas very rich in silica, such as rhyolite and dacite. The silica content of the steep peaks of San Francisco Mountain falls between that of the intermediate-sized mountains and the cinder cones—namely, andesite.

The expansive meadow at stop 2 provides an outstanding view west to San Francisco Mountain. The mountain has eroded into a

Two very different types of volcanoes, the San Francisco Peaks stratovolcano and the Sugarloaf lava dome

horseshoe-shaped ridge crowned by five high peaks that surround a central valley known as the Inner Basin, which you are looking up into. Among these five peaks are Arizona's three highest, collectively known as the San Francisco Peaks. A smaller, rounded dome squats at the mouth of the Inner Basin. This is Sugarloaf Mountain, which we will examine more closely at stop 4.

This vantage point gives you a good appreciation for San Francisco Mountain's immense size and steepness, both of which are due to the intermediate silica composition of its andesite lava. Andesite is a sticky, viscous material that cannot flow as readily as basalt, creating very steep slopes that typically rise to a sharp, central peak. Called stratovolcanoes because they consist of stratified layers of lava flows and volcanic ash, such mountains are most abundant along subduction zones. Famous examples include Mount Fuji in Japan and Mount Rainier in Washington State.

San Francisco Mountain used to be an elegant cone reminiscent of those iconic volcanoes, but not long ago (geologically speaking) it collapsed catastrophically to form the present horseshoe of smaller peaks. You can mentally reconstruct the volcano's original stature by tracing its gracefully curved ridges up in an ever-steepening arc to their junction at the former summit. Geologists performing more exact reconstructions have concluded that San Francisco Mountain once soared up to 15,000 or 16,000 feet—likely the tallest peak in the lower forty-eight states in its heyday.

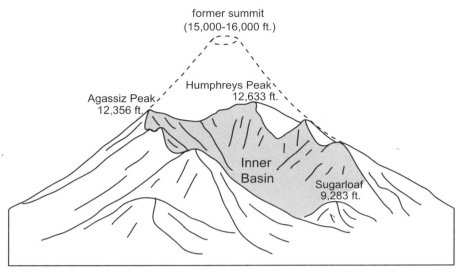

Prior to its collapse 400,000 to 200,000 years ago, San Francisco Mountain soared over 15,000 feet, making it the tallest peak in the lower forty-eight states at the time.

Such collapses are very common among stratovolcanoes. While their sticky andesitic lava allows them to grow large, it also lays the foundations for their inevitable demise. Unlike basalt, andesite does not allow expanding gases to escape easily, causing pressure to build up in the underlying magma chamber. Eventually, this pressure exceeds the strength of the overlying rocks, unleashing a violent ash eruption similar to the 1980 eruption of Mount Saint Helens. The ash blows miles into the air before settling on the volcano's flanks in thick, loosely consolidated layers. After the initial eruption depletes the magma chamber's store of gases, most stratovolcanoes ooze sticky, non-explosive lava that cements loose materials together. These "lava ribs" are strong enough to support steep slopes thousands of feet tall, but the loose ash layers contain structural flaws—ticking time bombs just waiting until some event triggers their failure.

So what event triggered the collapse of the mountain's upper walls and formed the Inner Basin? Geologists have come up with three different possibilities. Perhaps, after an enormous eruption, the original summit collapsed into the empty magma chamber, creating the Inner Basin. Just such an event formed Oregon's famous Crater Lake. A second alternative emerged after the eruption of Mount Saint Helens, which, prior to 1980, was a beautiful stratovolcano with a classic cone shape. The 1980 blast blew away its upper 1,300 feet, forming an inner basin ringed by jagged peaks strikingly similar to the configuration here. However, both of these scenarios are unlikely. Either of these proposed events would have showered northern Arizona with large volumes of volcanic ash, but no such deposits have been found.

A third possibility is that the mountain collapsed in a spectacular landslide unassociated with volcanic activity. Geologists studying Mount Rainier have found convincing evidence of several such slides that occurred on that mountain's steep slopes between eruptions. Because the Inner Basin of San Francisco Mountain opens out toward the northeast, if such a landslide occurred, the evidence would be right at your feet.

Before you depart to examine some of this evidence at stop 3, note how flat Bonito Meadow is. The topography in this area consists of either level surfaces like the meadow or rounded volcanic hills. When you go to stop 3, however, just a few hundred yards west, you'll see thickly forested countryside consisting of low, undulating hills. This hummocky topography may seem unremarkable, yet it is unusual. Unlike the hills to the east, no volcanic bedrock is visible on the flanks of these swales; they consist exclusively of loose debris. If you walk 50 yards south of the pullout, you will see that much of this loose material consists of cantaloupe-sized and larger blocks of reddish brown rock. This is andesite. None of the smaller volcanoes surrounding you produced andesite lava, so these blocks must have come from San Francisco Mountain. Their relatively large size required a high-energy process to transport them here. A

These cantaloupe- to watermelon-sized blocks of andesite at stop 3 were transported here during the collapse of San Francisco Mountain in a giant landslide.

sizable creek could have done the job, but there is no evidence of such a creek in the area. The blocks must have been carried here by some other means.

The hummocky terrain and large boulders both suggest that you are standing on top of a giant landslide. When geologists estimate the volume of material in this deposit, called the Sinagua formation, it closely matches the enormous volume of rock they estimate is missing from the top of San Francisco Mountain. The unassuming hummocks around you, it seems, could tell a tale of a monstrous landslide that decapitated the highest peak in the contiguous United States a fleetingly short time ago.

Radiometric dating indicates that the San Francisco Mountain volcano was built by repeated eruptions stretching from 1.8 million to about 400,000 years ago, so the mountain's collapse must have occurred more recently than that. Sugarloaf, the small, dome-shaped mountain that blocks the mouth of the Inner Basin, provides us with a minimum age of collapse. This small peak, which formed approximately 200,000 years ago, would only have had the room to form *after* the stratovolcano was gone. The landslide must have occurred between 400,000 and 200,000 years ago. Flagstaff's Museum of Northern Arizona has an excellent exhibit detailing the spectacular collapse of San Francisco Mountain. Don't miss this outstanding resource—northern Arizona's only natural history museum.

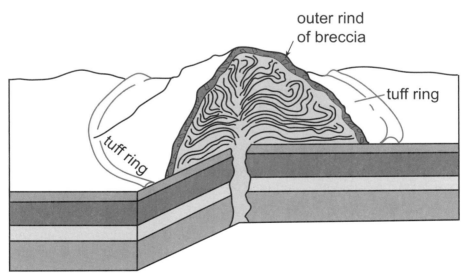

Lava domes like Sugarloaf form when sticky lava piles up like toothpaste. Hot lava flowing inside fragments the solidifying outer rind, forming volcanic breccia.

The short drive to stop 4 provides good views of Sugarloaf Mountain. Sugarloaf is a lava dome, a steep, loaf-shaped volcano that forms from the eruption of silica-rich lava. Like Sunset Crater and its Bonito lava flow, Sugarloaf formed in a two-stage eruptive process, beginning with a gas-rich explosive phase followed by a quieter flow of lava once the reservoir of gas was depleted. However, due to the very different silica contents of the magmas that fed Sugarloaf and Sunset Crater, the topographic manifestations of the two phases are completely reversed. At Sunset Crater, erupted cinders built the prominent volcanic cone during the explosive phase, while the valley-hugging Bonito lava flow formed when runny, basalt lava flowed from the volcano after the gas was depleted. For Sugarloaf, the dome formed during the later, nonexplosive phase, while the smaller hill of pumice you see at stop 4 was erupted during the earlier explosive phase.

The abandoned pumice quarry at stop 4 is a story in itself. Pumice mining is quite disruptive to the land—a price that was, until recently, being paid for that lived-in look in blue jeans, achieved by stonewashing with pumice. The San Francisco Peaks are sacred to the Navajo, Hopi, and other Native American tribes. When the quarry threatened to expand, the tribes put their foot down, and in fact began lobbying for the mine's closure. In 1998, Interior Secretary (and former Arizona Governor) Bruce Babbitt negotiated an agreement that shut down the mine out of respect for the tribes' religious beliefs.

You can easily observe the color of Sugarloaf's rocks here in the quarry. In stark contrast to the Bonito lava flow's black basalt, these rocks are almost white, indicating that they consist of rhyolite, a volcanic rock that contains over 70 percent silica. If you pick up a piece of pumice, you will notice its light weight. It is the only type of rock light enough to float. Pumice's extreme light weight is due to the large number of gas bubbles that were trapped as the magma solidified. The bubble cavities, or vesicles, are easily visible in the pumice here, especially through a magnifying glass. The vesicles testify to the very high gas content when the pumice erupted.

This quarry lies along a tuff ring, a low hill composed of pumice encircling a central depression that formed during a phreatic eruption. Phreatic eruptions occur when rising magma encounters abundant groundwater. The magma's heat flashes the water to steam. In the ensuing violent explosion, the magma, its heat quenched by water, transforms into pumice and flies in all directions, forming a tuff ring. The area directly above the explosion collapses, resulting in the depression. The tuff ring and its central depression are together referred to as a maar.

It is rare for an initial phreatic eruption to remove all of the magma in a chamber, so this explosive phase is commonly followed by a dome-building phase in which lava with very little dissolved gas is erupted. Such was the case here. In the absence of gas, incredibly sticky rhyolite lava filled the maar's central depression like toothpaste squeezed out of a tube, forming the 1,000-foot-tall Sugarloaf dome.

Lava domes usually grow by a process called inflation. Over time, lava is added to the dome's interior, forcing its rapidly solidifying skin to expand, much like an inflating balloon. Unlike a balloon, however, the lava's margins are not made of stretchy rubber. As the exterior cools, solidifies, and continues to expand, it simply shatters. Blocks break loose, tumbling down to form an apron of lava fragments. The resulting rubble covers the still-growing interior. Once the eruption is over and the dome solidifies, the inside is a solid mass of lava wrapped in a cloak of angular, broken rock called breccia. Unfortunately, due to the thick forest cover, Sugarloaf's mantle of breccia is hard to see, even close up.

If the gate near the quarry is open, drive up to stop 5 for a sweeping panoramic view to the east. From this lofty perch, you can see that Sunset Crater is the tallest of the small, dark cinder cones in this section of the San Francisco volcanic field. Another cluster of similar cones lies to your left (north); behind them is the North Rim of the Grand Canyon. O'Leary

O'Leary Peak, a complex lava dome. Note how much larger O'Leary Peak is than Sunset Crater, which is about as large as cinder cones can grow.

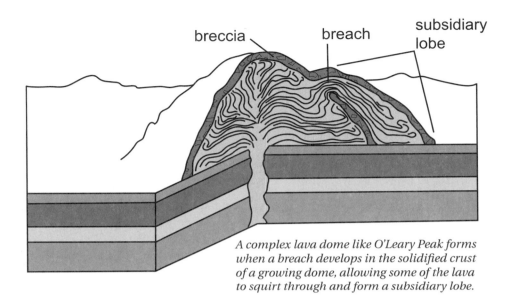

A complex lava dome like O'Leary Peak forms when a breach develops in the solidified crust of a growing dome, allowing some of the lava to squirt through and form a subsidiary lobe.

Peak, much larger, rises directly north (left) of Sunset Crater. Light-colored rock exposed in a quarry near O'Leary's base reveals that it, like Sugarloaf Mountain, is a volcanic dome composed of high-silica lava—in this case, dacite. Sugarloaf is a typical dome, small and conical. In contrast, O'Leary Peak is more sprawling and topographically complex. That's because, as the mountain ballooned during the eruption, cracks formed in its crust. Some of the magma came through these cracks, creating a new layer, or lobe, with its own exterior mantle of breccia, elongating the peak in a northeast-southwest direction.

Stop 5 lies at about 8,000 feet in a forest of ponderosa pines. From here, continue climbing the massive flanks of San Francisco Mountain to stop 6 at Lockett Meadow in the Inner Basin, tucked behind the west flank of Sugarloaf's lava dome. Here at 8,600 feet, the ponderosa pine forest gives way to a more diverse boreal forest that includes Engelmann spruce, Douglas fir (not a real fir tree), several varieties of true firs, and aspen trees. Lockett Meadow feels more like alpine Colorado or Wyoming than Arizona, but the massive size of San Francisco Mountain raises the landscape here high enough to bring a splash of Rocky Mountain vegetation to the Southwest.

A few hundred thousand years ago, San Francisco Mountain's flanks rose thousands of feet around you, culminating in a summit about 6,000 feet above your head. After the mountain collapsed and the Inner

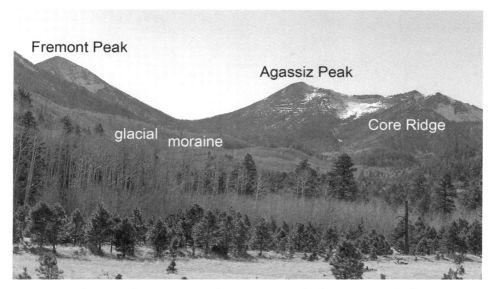

At Lockett Meadow, stop 6, in the Inner Basin, looking west to the former interior of the San Francisco Mountain volcano. Agassiz Peak is Arizona's second highest peak; Fremont Peak ranks third. An episode of glaciation in the valley between these peaks left a glacial moraine. Core Ridge is a remnant of the conduit that fed the volcano.

Basin formed, streams and glaciers during Quaternary time enlarged and reshaped this beautiful amphitheater. As you look west toward the highest peaks, you can see that the Inner Basin splits into two subbasins separated by a long, narrow ridge aimed straight at you. This is Core Ridge, a remnant of the conduit that fed magma to the volcano. When the mountain collapsed, the ridge's more resistant rock refused to yield, leaving it standing tall.

Today, snowmelt percolating into the mountain's porous volcanic rocks makes the Inner Basin a critical source of groundwater for the city of Flagstaff, which has tapped a number of natural springs here. The city has also drilled multiple wells and constructed a pipeline to deliver the water 2,500 feet downhill to the thirsty city in the stratovolcano's shadow.

Comparing and contrasting the volcanoes around Flagstaff raises a puzzling question. Why were so many different types of magma erupting in the San Francisco volcanic field? The answer lies in the way magma is generated. Deep in the earth, as a rock heats up and begins to melt, those minerals with the highest silica content melt first. Rocks rarely melt completely, and the magma that results from this partial melting process thus contains more silica than the rock from which it was derived. Less dense than the surrounding rock, the silica-rich magma rises toward the surface, separating from its now silica-poor parent rock like cream from milk. Partial melting of the upper mantle, which is exceptionally silica-poor (less than 45 percent), results in magma with slightly higher silica content, which produces basalt. The fact that the vast majority of the San Francisco volcanic field's peaks are basaltic cinder cones is testament to the fact that most of the magma fueling this field came from the mantle. The basalt magma must rise through the continental crust in order to erupt. The continental crust is much richer in silica, equivalent to andesite on our silica chart. Heat associated with this injection of rising basaltic magma can trigger partial melting of the more silica-rich continental crust, spawning subsidiary magma chambers richer in silica, and producing andesite, dacite, and rhyolite. Thus, the great variety of lava types found in the San Francisco volcanic field is the result of partial melting and other complex interactions between the rising magma bodies and the surrounding rocks of the continental crust.

The most unusual characteristic of northern Arizona's volcanism is that, unlike most volcanic fields, the eruptions here are occurring far from any tectonic plate boundary. A likely explanation is that Basin and Range extension, which has been migrating eastward across Arizona for the last 30 million years, is now impinging on the Colorado Plateau. Extension is frequently associated with a general heating and melting of the crust, which results in large-scale volcanism. The first eruptions are usually followed by normal faulting, which drops a block of crust down to form a broad valley and provides a conduit for even more magma to reach the surface. The nearby Verde Valley (vignettes 17 and 18), with its abundant volcanic rocks, is a great example of just such a faulted basin. It serves as a model of what Flagstaff might look like in the geologically near future.

East of Flagstaff, a nearly featureless plain stretches for miles along Interstate 40. That's how the area around Flagstaff could have looked. Instead, a myriad of obscure and seemingly insignificant factors, such as the amount of silica in a magma chamber millions of years ago, combined to create the mountain scenery around you, here on the roof of Arizona.

12
Desert Niagara
GRAND FALLS

If you are a first-time visitor to Grand Falls, nicknamed "the Niagara of the Desert," you may be unprepared for the scale of the spectacle before you. It's true that the Little Colorado River is no Niagara River. It conveys just a tiny fraction of that river's annual flow, and during many years, much of the Little Colorado is bone-dry for months at a time. If you visit Grand Falls during the early summer, you are likely to be met by a parched and uninspiring precipice barely distinguishable from the thousands of other cliffs lining the Colorado Plateau. But if you arrive when snowmelt or monsoon rains swell the stream and Grand Falls is in its full glory, you may come away feeling that the comparison to Niagara Falls is apt indeed.

If you are fortunate enough to see the Little Colorado River in full flood, you will witness over 20,000 cubic feet of water—150,000 gallons—pouring

Grand Falls in its full glory

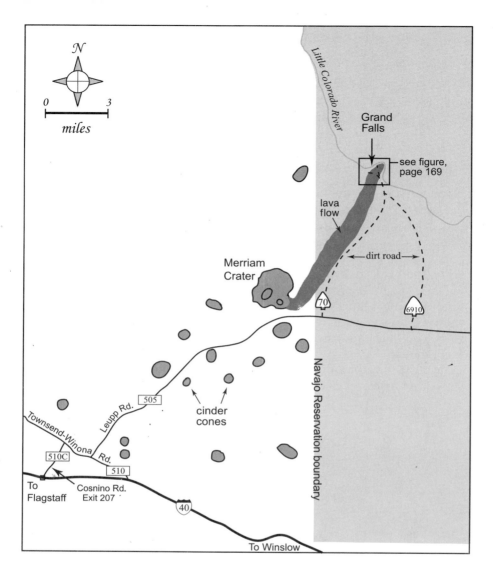

❖ GETTING THERE

From Flagstaff, drive east on I-40 and take Exit 207, Cosnino Road (County Road 510C). Turn left (north). Cross the railroad tracks, continue 2.2 miles, and turn right (east) at the T-junction on Townsend-Winona Road (County Road 510). Travel 1.9 miles, turn left on Leupp Road (County Road 505), and follow it northeast for 15.2 miles. Immediately after a sign welcoming you to the Navajo Reservation, turn left onto Navajo Route 70, a wide dirt road. Veer left at an unmarked junction at 3.6 miles and left again at a second junction with Road 6910 at 7.5 miles. After a total of 8.6 miles, turn left onto an unmarked, very bumpy dirt road for 0.4 mile. Park at any of the concrete sun shelters near the lip of the falls. (If you miss the last turn to the sun shelters, you will almost immediately reach a ford of the Little Colorado River.) The best times to visit Grand Falls are during the February–April snowmelt season and after a heavy rainstorm during the July–September monsoon season—both wet times, which can make dirt roads impassable, so proceed with caution (see preface).

over the falls every second, more than enough to produce a thunderous roar and launch a windblown spray worthy of its more famous cousin. Both falls are the same height, but unlike Niagara Falls, where the water plummets down a single, straight drop, Grand Falls surges down a series of stair steps, gracefully cascading to a final, dramatic plunge. In a flat, parched setting hosting few streams, this spectacle is perhaps even more stunning; such power and beauty, especially in the desert, is an infrequent and fleeting treat.

Grand Falls is a relative newcomer to the northern Arizona landscape. Although earlier estimates assigned it a much older age, new techniques date its formation to just 20,000 years ago, when the area looked topographically similar to what you see today.

As you drive toward the falls, you will drive mostly over hard, gray limestone that surfaces the majority of northern Arizona's level plains, including the rim of the Grand Canyon. Called the Kaibab formation, this rock was deposited in a warm, shallow sea that flooded the area 270 million years ago (see vignette 7). Occasionally you will see patches of bright red dirt, a colorful indication that you have climbed slightly higher in the rock stack, into the softer sandstones and mudstones of the overlying Moenkopi formation. These were deposited 25 million years later by sluggish rivers winding along a flat plain similar to today's lower Mississippi River valley. Slicing through these layers, the Little Colorado wanders through the area in its 200-foot-deep gorge, delivering water and sediments to its confluence with the main Colorado River in the easternmost Grand Canyon.

Poking up dramatically through this flat sedimentary veneer are a couple dozen steep, grass-covered hills made of heaped-up volcanic cinders. Called cinder cones, they are just a handful of several hundred dotting the landscape around Flagstaff, in what is known as the San Francisco volcanic field (vignette 11 discusses the history of volcanism there).

One cone in particular plays a leading role in the story of Grand Falls. This is Merriam Crater, the last large hill to your left as you drove along Leupp Road. Merriam Crater indirectly formed Grand Falls when a basalt lava flow erupted from a crack at its base. Basalt is a runny type of lava capable of traveling very long distances, and this flow surged northeastward for over 6 miles until it reached the gorge of the Little Colorado River. The dirt road you drove along followed the southeastern edge of this lava flow almost the entire way to Grand Falls. Sporadic clumps of black rocks lining the rim of the wash north of the road attest to the presence of this flow.

When the runny lava finally reached the gorge of the Little Colorado River, it spilled over the rim, tumbling down 200 feet in what must have

Merriam Crater. The lava flow that blocked the Little Colorado River issued from a fissure on the right flank of this cinder cone.

been a spectacular, fiery "lava fall." Once confined in the gorge, the basalt flowed downstream, partially filling the canyon for another 15 miles before grinding to a halt. Meanwhile, at Grand Falls, where the flow first met the river, so much lava poured into the gorge that it filled to the brim and then overtopped it, sending a tongue of lava surging across the channel another 0.5 mile to the northeast.

When the fireworks were over and the lava had solidified into dark volcanic rock, the Little Colorado found itself dammed by a basalt plug completely filling its former gorge. Because this plug blocked the river's flow, a lake began to form upstream. It didn't take long for the lake to reach the height of the lava dam and overflow, and when it did, the water followed the path of least resistance, making an end run around the toe of the basalt flow and plummeting back into its old channel over a limestone cliff that had, just a short time before, been part of its original canyon wall. Grand Falls was born!

How did geologists piece together this fantastic story? Because the falls formed so recently, the geologic evidence is still plainly visible, literally in black and white. When you reach the falls, there are a number of vista points both at the lip of the falls and, farther down the road, on the far (west) rim of the gorge directly opposite the falls. If you walk or drive to a vista point across from the falls and look at the southeastern wall (to

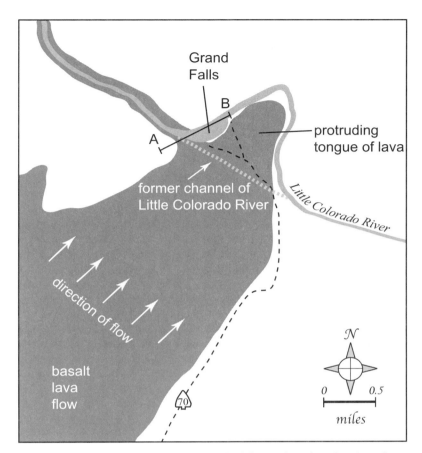

The lava flow filled the former channel of the Little Colorado River, then flowed downstream for another 15 miles. Where the lava filled the canyon, a tongue of basalt pushed across the gorge to the other side. This basalt plug diverted the Little Colorado River and created Grand Falls where the river spilled back into its gorge. Line A–B marks the location of the diagrams on page 172.

your right as you face the falls), you can see a perfect cross-section of the gorge as it appeared before the volcanic chaos began. First focus your attention on the sun shelter that rests on a knob of dark basalt directly above the lip of the falls. As you trace down the cliff below this shelter, you will notice that the black basalt is relatively thin and gives way to the brownish-white Kaibab formation just 20 feet below. Now scan across the cliff further to the right. Notice that here the wall is composed of

The geologic evidence for how Grand Falls formed is still plainly visible because erosion has revealed a section of the Little Colorado River's former gorge, now filled with dark basalt, beneath the right-hand sun shelter. The left sun shelter rests atop the thin basalt tongue that overtopped the gorge.

basalt from rim to river, a full 200 feet thick. What could account for this dramatic difference in thickness? You are actually looking at the old river gorge where it was totally filled with basalt—the very spot where Merriam Crater's lava poured over the rim and completely blocked the river. The thinner section of basalt to the left of the old channel is the tongue of the flow that overtopped the original gorge and spilled over the far wall. Now, look to see which type of rock you are standing on. This, too, is black basalt, and its 200-foot thickness here indicates that you are actually standing on top of the old lava dam. The Little Colorado River used to run right under your feet before the lava flow turned the topography upside down. Similar lava dams have repeatedly blocked the Colorado River in the western Grand Canyon—at least thirteen of them over the last 640,000 years (vignette 2).

In addition to the old lava plug, as you look downstream you can still see remnants of the basalt that flowed down the river canyon and partially filled it. Once the Little Colorado River regained its original gorge by forming Grand Falls, it immediately set to work slicing through the basalt that had so callously blocked its path. The dark rock you see lining the downstream walls reveals where the river has cut right through the old basalt flow, slowly but surely wearing it away and exposing these remains.

When you look at the modern river gorge, the walls are very steep. Since the old gorge was carved in the same rock, the brand-new waterfall most likely featured a steep, Niagara-like plunge. How, then, did Grand Falls evolve into the signature cascades that we see today? The "quick and dirty" answer is just that: dirt. Grand Falls' transformation depended on the sediments hauled by the river water, as well as the characteristics of the limestone over which it plunged.

If the river is flowing, check to see what color the water is. Instead of brilliant blue, the water is usually the color of chocolate milk. This is because the river runs thick and brown with sediment that it greedily devours during its transit across the Painted Desert, the vivid landscape that is sculpted from the extremely soft and easily eroded Chinle formation (the rock layer that hosts the impressive deposits of petrified wood discussed in vignette 15). Every year, the Little Colorado River hauls 8.5 million metric tons of sediment downstream—nearly 567,000 dump truck loads. This is a fabulously large amount for so small a river to carry, and it gives the Little Colorado sandpaper-like erosive power.

Immediately after Grand Falls formed, the abrasive action of the Little Colorado's sediment-laden water began to gnaw away at it. Water flows more swiftly where the river gradient is very steep—for example, at rapids and waterfalls. The swifter and more turbulent flow translates into accelerated erosion at these steep areas, known as knick points. The turbulence at the base of waterfalls quickly erodes the lowermost rock. If you are lucky enough to be here when the Little Colorado is in flood, look toward the bottom of the falls, where an enormous eddy forms. There, dozens of logs as well as thousands of tons of sediment swept down by the flood slowly churn and circulate in the water, banging forcefully against one another and the rock. This erosive action will eventually undercut the entire cliff, triggering landslides that will shift it, and consequently the waterfall, upstream over time.

It is this upstream migration in the Kaibab cliff that formed Grand Falls' distinctive cascades. Migration progresses at different speeds on different sedimentary bedding layers, depending on their relative strength. In the Kaibab formation, harder, more limestone-rich layers erode more slowly and tend to form steep steps, while softer, sandier layers erode more quickly. The staircase of cascades and falls you are marveling at

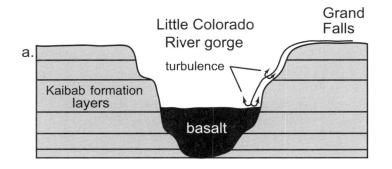
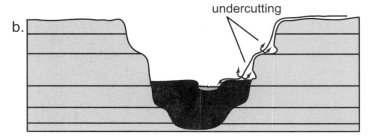
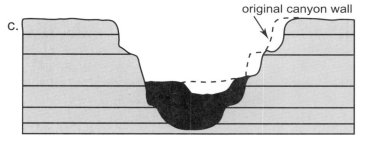

Turbulence at a waterfall progressively undercuts the cliff face (shown in panels a through e), which collapses in a landslide when it loses this support. Because of the layered nature of the Kaibab formation, Grand Falls has undercut several separate strong layers, creating a series of smaller falls instead of one larger one.

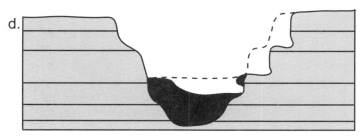
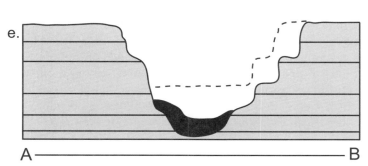

is the result. Had Grand Falls poured over a more uniform rock, such as basalt, it would likely have remained a single, Niagara-like drop.

Ultimately, through this upstream migration, Grand Falls will disappear completely. As a waterfall shifts slowly upstream, the river bottom lies progressively higher, so the falls become shorter. Fundamental physics dictates that all knick points follow a similar evolutionary path, migrating upstream and becoming fainter until they vanish completely. Niagara Falls is currently shifting upstream at an average rate of about 1 foot per year; in the past, that rate was even faster, and it now lies over 7 miles upstream from where it first formed about 12,500 years ago. Grand Falls is beating a similar retreat. The fact that the falls remain so close to the original canyon wall is a testament to its extreme youth.

Besides helping to carve Grand Falls, the Little Colorado's heavy sediment load is important for another reason. The 8.5 million metric tons that pour over the lip of the falls each year eventually make their way to the Grand Canyon, where the sediment swirling below your feet has achieved great ecological significance since 1963. This is the year Glen Canyon Dam was completed, trapping most of the 57 million tons of sediment that used to flow down the main Colorado River through the Grand Canyon (vignette 3).

The Canyon's predam ecosystem relied on organic matter swept along with that sediment, and its absence has radically disrupted the base of the food chain. Organisms that once derived their energy from the organic matter have largely perished, replaced by photosynthetic organisms that thrive in the clear waters released from the dam. The reduced load of sediment moving through the Grand Canyon has also impacted sandbars and beaches along the riverbanks. Prior to 1963, large quantities of sand moved downstream during floods. As each flood waned, the sand dropped to the riverbed, building up bars and beaches. These provided important habitat and protected archaeological artifacts left behind by the Ancestral Puebloan people (formerly known as the Anasazi).

Without the tons of sediment now trapped behind Glen Canyon Dam, the remnants of the native aquatic ecosystem rely on the sand and organic detritus delivered by just two tributaries. One is this river, the Little Colorado. Scientists are trying to take full advantage of its precious cargo to rebuild Grand Canyon beaches and halt, or at least slow, the demise of several of its endangered species.

To get a tangible sense of just how much sand the Little Colorado River carries, and also for a stunningly different view, you can take a short hike down to the base of the falls. From the last wide dirt parking area (the one farthest downstream) on the canyon rim road, a faint trail leads down to the river. The steep path threads its way between the rim and a prominent tower of basalt, a remnant of the Merriam Crater lava flow that has not yet been toppled by the river's abrasive caress. At the bottom, you will see in the basalt a series of parallel, vertical cracks. These are cooling joints, which formed as the lava quickly cooled and contracted. Carefully make your way across the slippery rocks to the river. To see for yourself how much sediment the river is transporting, fill up a glass or water bottle and let the sediment settle out. The large amount you will capture if the flow is high will graphically illustrate how quickly Arizona's Painted Desert is being ripped up and carted away.

Though Grand Falls' sporadic flows and dirt access road keep it from being the famous tourist attraction that Niagara Falls became, that very remoteness and ephemerality, along with the falls' fiery history, make a backcountry detour to Arizona's "Niagara of the Desert" more than worthwhile.

13

Pangaean Riviera
THE SEDONA RED ROCKS

Sedona is routinely hailed as one of the most beautiful places in America. Tourists, retirees, and hikers from around the globe are drawn to the enchanting landscape of soaring red spires rising abruptly from soft hills clad in pine, juniper, and cedar. Practitioners of New Age religions claim that the area holds many vortices—spots where, it is believed, the earth's power emanates from the ground.

The area's stunning beauty is primarily due to geologic events that transpired here from 280 to 270 million years ago, during late Permian time, when the region was perched at the edge of a gigantic landmass, or supercontinent, known as Pangaea. Because Pangaea was an amalgamation of all the present-day continents, coastal real estate was a rare commodity back then. The bit of coastline that existed was sun-drenched and lined with miles of pristine, sandy beaches—a regular Pangaean Riviera! This low-lying area experienced 10 million years of nearly continuous deposition of sediments, later cemented into rock and carved

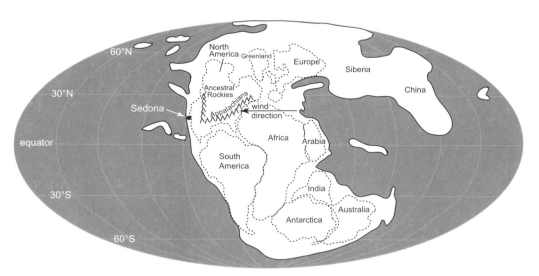

Map of Pangaea during Permian time. Sedona lay on the arid west coast of this supercontinent, cut off from moisture by the vast continent upwind and the storm-blocking heights of the youthful Appalachian and Ancestral Rocky Mountain ranges.

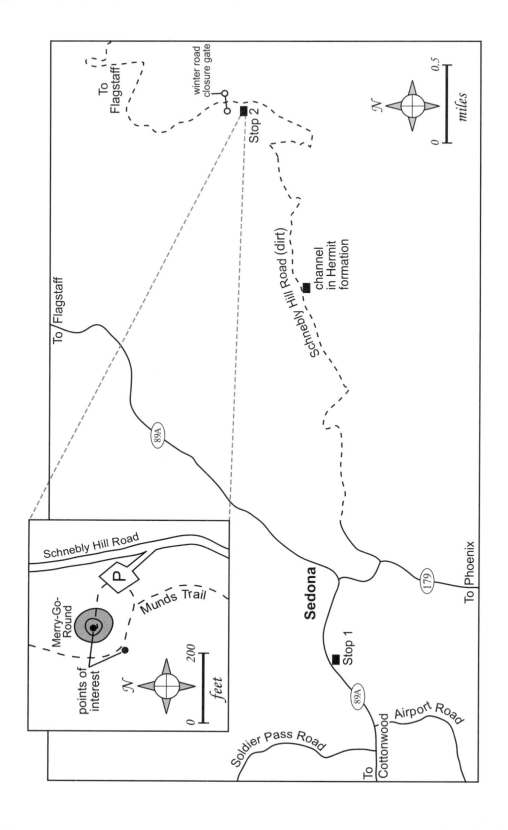

into modern Sedona's red walls and spires. Subtle changes in deposition recorded in these rocks allow us to journey back to Permian time and trace the interactions of land and water along this long-vanished coastline. We can also observe firsthand how these variations have influenced the modern landscape.

Beautiful red rock towers of Sedona, such as Cathedral Rock (right), *are composed of the Schnebly Hill formation. The Schnebly Hill has three members: the Bell Rock, the Fort Apache, and the Sycamore Pass. Below these, the Hermit formation floors the valley; the overlying Coconino sandstone caps the highest mesas.*

GETTING THERE

Sedona lies 27 miles south of Flagstaff at the intersection of Arizona 89A and Arizona 179. From this intersection, head west on Arizona 89A for 0.7 mile and look for a low roadcut, stop 1, on the south side of the road; park in the lot across the road. To get to stop 2, return to the intersection of Arizona 89A and Arizona 179. Turn right (south) onto highway 179 and go 0.3 mile to a clearly marked left (east) turn onto Schnebly Hill Road, which soon turns to dirt. After 2.1 miles, you pass on the right a large channel in the Hermit formation similar to the one you saw at stop 1. Continue an additional 2.6 miles to stop 2, a small parking area on the left just past milepost 4 and just before a metal gate. The Munds Trail has an initial short downhill stretch followed by flat walking. The trail up to the Merry-Go-Round formation is an easy five-minute walk.

The Hermit formation was deposited in a series of river channels and adjacent flat floodplains. The scalloped nature of the beds in this outcrop at stop 1 reveals an ancient channel that crossed the area 280 million years ago.

Three main rock layers comprise Sedona and its environs. Most of Sedona is built on the soft, brick-red Hermit formation. The red rock walls and spires rising above the tourist shops, luxury homes, and New Age bookstores are carved from the Schnebly Hill formation, and the distinctive white caps worn by the tallest walls are Coconino sandstone.

Our journey back to the Permian begins at the roadcut at stop 1, hewn from the Hermit formation. In the cut, maroon sandstone alternates with reddish purple mudstone. Note the layers' scalloped nature. One group of beds cuts a rakish swoosh through the outcrop, truncating another set, only to be itself severed by the arched layers of another group. These packages of sedimentary rock, called lenses because of their eyelike shape, demarcate ancient river channels that crisscrossed the Sedona area about 280 million years ago. The sand that makes up this sandstone accumulated in these channels, whereas the mudstone's soft mud blanketed adjacent floodplains when the rivers overtopped their banks. Today, the area's soft green slopes, which contrast with the steep, bare rock walls above, are formed on these floodplain mudstones. Like modern meandering rivers, these ancient river channels gradually shifted back and forth across the floodplain. As the swifter current on the outside of bends caused the rivers to relentlessly gnaw away at these areas, slacker currents on the inside of meanders dropped sand, filling in those portions of the channels. This shifting created the complex stack of lens-shaped channel deposits you see here.

The geometry of these and other small channels throughout the Sedona area resembles that of the modern, ephemerally flowing arroyos that abound today in Arizona's dry climate, suggesting that the Sedona of 280 million years ago was similarly dry. Nodules of ancient caliche support this notion. Caliche, a form of limestone, develops in arid regions when groundwater charged with calcium and carbonate is wicked toward the surface by the unrelentingly hot sun. As the water evaporates, the calcium and carbonate bond to form these hard lumps. Fossilized impressions of extinct desert plants reinforce this interpretation.

The Hermit formation is extensively preserved across northern Arizona. In the Grand Canyon, it displays features similar to those you see here (vignette 7). Age-equivalent rocks in northeastern Arizona known as the Organ Rock shale (vignette 6) possess nearly identical characteristics, indicating that all of northern Arizona at that time was a vast, flat, arid plain traversed by a few large, perennial rivers that sluggishly wandered toward a sea a short distance west of here. Farther northeast, in Utah and Colorado, rocks of identical age contain increasingly bigger particles: first pebbles, and then, east of Moab, cobbles. This increasing size indicates mountainous country where swifter water currents can transport bigger material. Finally, just east of the Utah-Colorado border, granite boulders the size of economy cars, which even the steepest of mountain rivers could not transport, suggest talus from an extremely steep mountain range—the source of the Hermit formation sediments. This range is one of several that ran northwest-southeast through Colorado and New Mexico. These ranges rose in the same vicinity as the modern Rockies and are known collectively as the Ancestral Rockies. They were heaved up 300 million years ago by the collision of northern South America with Texas during the assembly of Pangaea.

The presence of the Ancestral Rockies reveals not just the source of the material that filled the Hermit formation channels here in Sedona as the ranges slowly eroded; it also explains the region's climate at that time. Sedona was then situated in tropical latitudes just north of the equator near Pangaea's west coast. In the tropics, prevailing winds blow from east to west, and it was no different 280 million years ago. Moisture in these winds was progressively wrung out on their long journey across the continent toward Sedona. Any remaining drops were effectively removed as the air masses trudged up the Ancestral Rockies' steep slopes. Sedona and most of the modern American Southwest lay parched in the range's rain shadow.

Ultimately, this ancient desert climate is responsible for the vivid red color of the Hermit formation, the overlying Schnebly Hill formation, and many other beautiful layers in the American Southwest. Granites like those shed off the Ancestral Rockies are rich in iron-bearing minerals. When exposed to the atmosphere, these minerals break down chemically, forming a series of iron oxide compounds ranging in color from red to mustard-yellow to brown. In arid climates the red iron oxide mineral

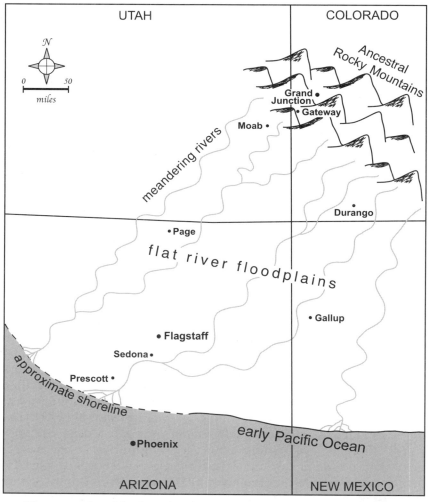

The geography of the Four Corners area when the Hermit formation was deposited about 280 million years ago. At that time, Sedona lay near the terminus of rivers that drained off the Ancestral Rockies.

known as hematite is especially abundant. Hematite is the stuff of common rust, and the rocks of the Ancestral Rockies literally rusted on their long journey to the sea. The sand in the Hermit channels is 99.5 percent quartz, a colorless mineral. But every one of these quartz grains is coated with hematite dust, staining the whole formation a deep, dark red—analogous to a few drops of food coloring in a batch of dough.

Shortly after 280 million years ago, the few perennial rivers of the Hermit formation dried up, and no sediment accumulated across this area for the next couple of million years. Why the rivers disappeared is not completely understood, but fortunately for our tale, sedimentation soon resumed. The rocks that record this sequel are those of the Schnebly Hill formation, wonderfully exposed along Schnebly Hill Road, one of the most beautiful drives in the entire Sedona area. As you head to stop 2, pause at the much larger Hermit formation channel on the way, likely formed by one of the perennial Permian rivers. Look for the telltale lens shape similar to those at the first stop, only much larger.

Stop 2 is the Merry-Go-Round, a rock formation resembling its namesake. From the parking area walk 150 yards west down a gentle gully floored with slickrock to a flat, prominent bench. Here you join up with the signed Munds Trail, marked by huge cairns held together with chicken wire. Once you reach the trail, veer right. Turn left after the second cairn and walk 20 yards to a promontory on the edge of a small escarpment. The view down the canyon toward Sedona is breathtaking. Now take a look at the rock on which you're standing. Its gray, pockmarked, resistant limestone top forms the flat floor of the Merry-Go-Round, encircling the small knob of red sandstone immediately above you.

The gray limestone and the red sandstone both belong to the Schnebly Hill formation. A formation—the fundamental unit by which geologists classify sedimentary rocks—is a series of layers sharing at least one characteristic distinctive from surrounding strata. Around Sedona, red sandstone cliffs characterize the 700-foot-thick Schnebly Hill formation lying above the slope-forming mudstones of the Hermit formation and below the towering white-to-tan Coconino sandstone cliffs. The Schnebly Hill contains four members. One is restricted to a localized area south of Sedona, but the other three are widespread and regionally significant. From where you stand, you can see all three. The most distinctive is the gray limestone band upon which you're standing, the Fort Apache member. As you look southeast along the escarpment, you can see that it is only about 8 feet thick. Below it, red sandstone exhibits prominent horizontal layering. Known as the Bell Rock member, this subunit was named for a famous landmark (and site of one of Sedona's reputed vortices) south of town. The knob of red rock above you is part of the highest

The blocky, light-colored Fort Apache member of the Schnebly Hill formation caps and protects the softer, horizontally bedded sandstones of the Bell Rock member below. Erosion of the soft sandstone undercut some of the limestone, forming a small natural arch.

member, the Sycamore Pass member. At first, it looks identical to the Bell Rock member, but we will examine some important differences shortly.

From your perch, you can trace the Fort Apache limestone across the cliffs far to the northwest; it forms the blocky, gray, protruding band girdling the entire wall about halfway up. The horizontal banding of the Bell Rock member is visible below the limestone, while above it the Sycamore Pass member's steeper red cliffs rise another 200 to 300 feet before giving way to the blond Coconino sandstone. The alternation between red and blond layers in these upper cliffs suggests that the transition from the Sycamore Pass member to the Coconino is gradual.

Look carefully below you at the Bell Rock member's prominent horizontal beds. Their sand and silt grains are weakly cemented, causing the Bell Rock to crumble out from under the Fort Apache, in the process creating a small arch capped by the more resistant limestone. Such monotonously horizontal layering is often created when sediments are washed back and forth in a nearly flat tidal zone at the ocean's edge.

Looking north from stop 2, the Fort Apache limestone member forms the prominent bench, with the horizontal layers of the Bell Rock member below. The cliffs rising above the bench consist of Sycamore Pass sandstones, in turn overlain by lighter-colored cliffs of Coconino sandstone toward the top.

Further corroborating this, symmetrical ripple marks occur throughout the unit. While currents flowing in one direction, such as in a river, produce ripple marks that are steeper on the downstream side, bidirectional currents like those on a beach or in a tidal flat form ripple marks that are symmetrical.

Tidal-flat deposits are usually composed of mud, not the sand you see here in the Bell Rock member. Scientists examining Bell Rock member sand grains under a microscope have managed to solve this puzzle. High magnification reveals frosting and pits on each grain, similar to glass that has been sandblasted. This is a distinctive trait of windblown sand grains, which, during transport, constantly collide at high velocity. In contrast, water molecules cushion waterborne sand grains. The collisions aren't strong enough to frost grains, they only round off their sharp edges. The frosting of Bell Rock sand grains means that, although ocean tides transported them to their final resting place, it was the wind that brought them to the water's edge. The tidal flats that occupied Sedona 278 million years

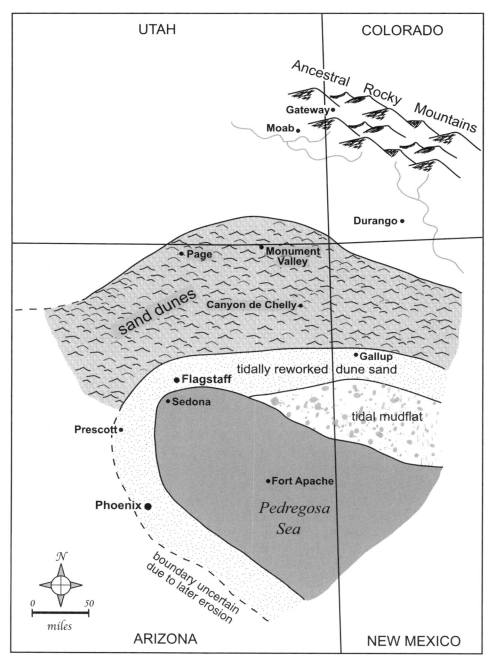

The Sedona area lay at the edge of the Pedregosa Sea 277 million years ago. Sand dunes covered the low-lying, arid land to the north.

ago were flanked by massive desert dunes that marched right up to the edge of the sea.

Shorelines shift according to the raising and lowering of both land surface and sea level. Since the deposition of the Hermit formation on a low river floodplain a few million years earlier, the Pangaean shoreline had risen relative to the low-lying plain. Characteristics of the Bell Rock member tell us that the Sedona area now lay right at the shoreline—so if you put your finger on the contact between the Bell Rock and Fort Apache members, you are touching the edge of a supercontinent. As the sea encroached upon the land, the Pangaean Riviera was soon inundated by a shallow sea, called the Pedregosa Sea, which deposited the limestone of the Fort Apache member about 277 million years ago. Soon after, the sea drained back out of the Sedona area after only the briefest of incursions, explaining why the limestone layer is a scant 8 feet thick here. In contrast, toward the southeast, the Fort Apache member progressively thickens to over 100 feet. Here the sea was deeper and lasted longer. Conversely, if you trace the Fort Apache member to the northwest, it becomes thinner and sandier until it disappears altogether in the Boynton Canyon area

In the Sycamore Pass member, along the trail to the top of the Merry-Go-Round at stop 2, crossbeds (tilted layers) terminate against horizontal layers.

A hill with the Sycamore Pass member at its base. The increase in number and thickness of the light-colored sandstone layers (and reduction in the horizontal, tree-covered bands) as you move higher indicates a general fall of sea level during this time. At the prominent bench halfway up the mountain, the red sandstones disappear entirely, marking the base of the Coconino sandstone, formed in a giant desert after the sea retreated even farther.

just west of Sedona. Sedona thus marks the zenith of the Pedregosa Sea's shoreline. The land adjacent to that sea was a desert covered with sand dunes, now preserved in the DeChelly sandstone found in Monument Valley (vignette 6) and at Canyon de Chelly (vignette 9).

As the water receded to the southeast, the sand dunes, which had never been far away, once again encroached upon Sedona, depositing the 276-million-year-old Sycamore Pass member. To examine this rock, return to the parking lot and take the wide trail up the eastern flank of the rounded knob above the limestone shelf. Though this subunit looks like the Bell Rock member, a detailed examination of its layering reveals important differences. In some places along the trail, the Sycamore Pass member consists of sandstone mixed with siltstone that is horizontally layered, making it identical to the Bell Rock layers below the limestone. Along other sections of the trail, however, you will see crossbedding (more about crossbedding in vignette 4). The crossbeds, composed entirely of sand, are quite large and fairly steep, indicating deposition in a desert dune. Microscopic examination shows that grains in these beds have the frosted, pitted surfaces diagnostic of wind transport, sealing the interpretation.

At the top of the knob, look northwest for a gorgeous view of walls and towers. The gray, blocky band of the Fort Apache member in the escarpment provides a convenient marker bed by which to orient yourself. Above it, the Sycamore Pass member holds sections of horizontal layering interspersed with crossbedded sections. The packages of crossbedding become thicker and more numerous higher up the walls, particularly in the zone of alternating red and white bands marking the transition between the Schnebly Hill formation and the overlying Coconino sandstone.

Subtle changes occurring along the Pangaean Riviera caused these differences. While the sediments of the Sycamore Pass member were deposited, sea level fluctuated slightly, but the desert dunes were never far away. When sea level stood a bit higher, a tidal flat covered the Sedona area. Windblown sand was swished back and forth and mixed with silt until it was laid down in horizontal sheets, just like the Bell Rock sediments a few million years earlier. At times when sea level dropped slightly, the tidal flat shifted southeast, and coastal dunes covered the landscape, leaving crossbeds behind. Another slight rise in sea level, and the tidal flats were back. On your hike up the knob, you pass several such alternations between horizontal planes and crossbeds. Especially limy horizontal layers at the top of the knob suggest deposition during a higher ocean incursion. Eventually, farther up the cliff walls (higher than where you are at the top of the knob), horizontal layers cease altogether, replaced by crossbeds deposited as desert dunes enveloped the Sedona area.

North and west of Sedona, the Schnebly Hill formation is absent altogether. In the Grand Canyon, the Coconino sandstone lies directly on top of the Hermit formation (vignette 7); their contact represents a 5-million-year gap in time. The absence of the Schnebly Hill there indicates that during this period, the Grand Canyon region was an upland area, a source of eroding sediment rather than a zone of deposition. The consistent southward tilt of the Schnebly Hill crossbeds, which are deposited on dunes' downwind sides, suggest that 276 million years ago, north winds carried sands from the Grand Canyon to form the magnificent towers around you.

The highest red layer on the canyon walls to the northwest marks the last gasp of the Pedregosa Sea. The blond Coconino sandstone above consists exclusively of coarse-grained sandstone. Its crossbeds are tilted at 30 degrees—the angle of repose, the steepest angle at which a pile of loose sand can slope—indicating an interior desert with huge dunes. The surfaces of some Coconino crossbeds contain tracks of spiders, scorpions, and rodentlike reptiles, confirming that these rocks were laid down on land. The 500-foot thickness of the Coconino in the taller cliffs to the south show this was a long-lasting desert that persisted from 275 to 272 million years ago.

Bouncing back down Schnebly Hill Road toward Sedona, you pass rocks whose story you have now unraveled. In them is recorded the complex interplay between sand and sea along this ancient Pangaean Riviera—an interplay without which the scene around you would never have formed.

14
A Crater with Deep Impact
METEOR CRATER

Fifty thousand years ago, a quiet day on the northeastern Arizona plains was shattered by the deafening roar of many cubic miles of air struggling to dodge a 300,000-ton ball of iron one-third the size of a football field that was traveling at forty times the speed of sound—over 26,000 miles an hour. A moment later this extraterrestrial wrecking ball and a huge mass of air that had not slipped out of its path slammed into the ground. Within 3 to 4 seconds, the vast majority of the colliding meteor had vaporized; 175 million tons of rock had been excavated to form the crater; and winds in excess of 1,200 miles an hour had leveled every tree within 12 miles. Two hundred times more powerful than the atomic bombs dropped on Hiroshima and Nagasaki during World War II, the 2.5-megaton explosion shot debris high into the atmosphere, blotting out the sun and raining down pulverized bits of space rock over northern Arizona for several days. When the cataclysm was over, virtually every living thing within several miles of ground zero was dead, and a crater three quarters of a mile across and 700 feet deep had replaced a featureless plain.

The plants and animals of the region, including pine trees, mammoths, mastodons, giant sloths, and bison, quickly repopulated the devastated

Meteor Crater from the air
—Courtesy Meteor Crater Enterprises

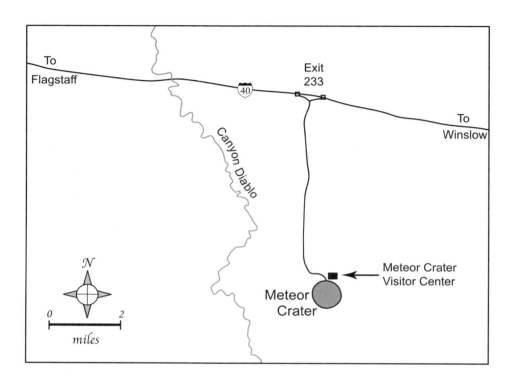

❖ GETTING THERE

Meteor Crater is 35 miles east of Flagstaff and 20 miles west of Winslow. Get there by taking exit 233 off I-40. A paved road leads 6 miles south to the privately owned crater, visitor center, and museum. Crater viewpoints are accessed through the visitor center. You may want to bring along a compass.

area. A lake filled the new crater, a welcome oasis for migratory birds. Over millennia, the climate warmed and dried out; many life forms disappeared or retreated to higher ground. The grasses that today cover the area moved in, and the crater filled with sediments, shrinking its depth from 700 feet to its current 594. As people migrated into the surrounding plains a few thousand years ago, they picked up fragments of the meteorite that littered the ground. Clearly they considered them significant; in two cases, archaeologists have found meteorite pieces buried in graves identical to those used for people. One fragment was even interred as children were, wrapped in special cloth made with feathers.

Thousands of years later, the crater and its fragments continue to fascinate humans. Study of the crater, beginning around 1900, impacted the science of the time and every decade since. We have learned that the earth is not an island, but rather interacts with the solar system in profoundly important ways.

As you approach Meteor Crater from I-40, you first see a raised ridge oddly out of place in an otherwise flat landscape. Passing through the visitor center and viewing the crater from the rim, it is difficult to grasp the scope of this hole in the ground. For a sense of scale, use the viewing scopes at various lookout points to locate the life-size plywood astronaut in the center of the crater's floor. In fact, the Apollo astronauts underwent scientific training at Meteor Crater in the 1960s since it was the only confirmed impact crater in the world. The fascinating Meteor Crater museum has detailed exhibits about the astronauts' work here and much, much more. During your visit, leave time for the guided rim walk, which provides great geological and human history—and includes beer cans the astronauts left behind.

Although Meteor Crater had been a well-known geographic feature for decades, its role in a scientific revolution didn't begin until one spring day in 1891 when a prospector in nearby Canyon Diablo picked up one of the hundreds of heavy metallic rocks littering the ground. The rock disappointed the prospector, who had hoped for a fortune, but found its way to a Philadelphia mineral dealer who recognized it as a meteorite—an extraterrestrial rock that enters earth's atmosphere and lands on the surface. Scientists who heard of the discovery immediately recognized that the crater could inform a debate then raging about how the craters on the moon had formed. Many scientists argued for lunar volcanism, but a second camp suggested that collisions with asteroids—rocks from space—had formed the craters.

Grove Karl Gilbert was one of the preeminent geologists of the time. In the summer of 1892, Gilbert went to Meteor Crater to determine whether an impact event or a volcanic eruption had formed it. Gilbert was a firm

The meteorite wreaked havoc on the pine-covered plains of northern Arizona, killing everything within several miles, including mastodons, sloths, and camels.
—Dona Abbott

believer in the asteroid impact hypothesis, but he was also a careful scientist who demanded that the facts on the ground fit his theory. Based on the theory of the time, Gilbert believed that the meteor must have been nearly as large as the crater itself and that, during impact, it had lodged itself beneath the crater floor. This theoretical bias led to two assumptions he would test. First, if the Canyon Diablo meteorites, composed of a highly magnetic iron-nickel alloy, were fragments of the larger rock, then compasses would go haywire as he approached the impact site. Second, if the meteorite lodged beneath the crater was the same size as the crater, then the amount of material ejected during the impact, which must now blanket the adjacent area, would be about twice the volume of the hole itself, consisting of material displaced from both the crater and from the area beneath the crater now occupied by the meteorite. Therefore his second test was to calculate the volume of the crater and compare this to the volume of debris covering the nearby landscape.

When you visit the crater, you can perform the same tests Gilbert did. If you brought along a compass, pull it out. Does it behave normally or go haywire? Gilbert's compass behaved completely normally. No large body of iron lay beneath the ground here. What about Gilbert's second test? As you drove south from the interstate toward the crater, you passed a number of layered, brick-red outcrops. This is the Moenkopi formation from Triassic time, the area's uppermost bedrock. It is not chaotically jumbled debris, as you would expect from an impact. In fact, the debris blanket here is quite thin. The crater's rim rises only 100 to 200 feet above the surrounding plain, and debris accounts for less than 100 feet of that height. If you imagine shoveling that debris back into the hole, you see there is too little to fill the crater, never mind twice its volume. Compiling more precise results, Gilbert reached the same conclusion.

Based on his tests, Gilbert rejected the impact hypothesis for the origin of Meteor Crater. He also saw that no volcanic rocks exist in or around the hole, casting doubt on the volcanic hypothesis as well. Gilbert declared Meteor Crater the result of a phreatic eruption—an eruption that involves superheated steam instead of lava—packed his bags, and headed home. For orthodox geologists, Gilbert's work closed the book on the crater's origin. The lunar crater debate continued to simmer, but Meteor Crater was viewed as unrelated.

But in 1902 a mining engineer named Daniel Barringer heard about the crater. Sure it could only be explained by a meteor impact, Barringer shared Gilbert's beliefs that the meteorite must be the same size as the hole and must lie buried beneath it. He secured mining claims to the crater, intent on exploiting this ore from the heavens. For twenty-seven years, Barringer prospected for the elusive meteorite, drilling twenty-eight holes up to 825 feet below the crater floor, but none ever penetrated the iron meteorite. With the visitor center's viewing telescopes, you can see a boiler, sheds, and other remnants of his mining operation at the bottom of the crater, as well as a supply road and waste-rock piles to the south. These piles are from slanting shafts he drilled after concluding the meteorite must have come in at an angle. Of course, these also failed to produce a giant meteorite. Barringer's investors eventually abandoned him, and he died of a heart attack soon after.

Although Barringer never found the meteorite or his fortune, he amassed a significant body of evidence supporting Meteor Crater's extraterrestrial origin. His drill shafts penetrated a 35-foot-thick layer of debris below the crater floor that was graded from large particles at the bottom to tiny particles at the top. This conformed to what one would expect from a rain of debris shot violently into the air, with the largest fragments falling out first, and smaller, lighter particles settling afterward. Below

Shafts drilled in the south crater wall by Daniel Barringer in his relentless search for the iron meteorite. The piles of light-colored debris two-thirds of the way down the crater wall reveal the shafts' location.

the debris layer, Barringer found 600 feet of highly fractured bedrock, seemingly crushed by a mighty hammer blow. Significantly, though, the bedrock below that was pristine and undisturbed—an argument against a phreatic explosion from below.

Intimately familiar with the area's stratigraphy after years of prospecting, Barringer also noted an important clue in the crater's sequence of rock layers. Meteor Crater is gouged into late Paleozoic through early Mesozoic layers typical of this part of the Colorado Plateau. The brick-red Moenkopi formation rock you observed on the drive in was deposited about 245 million years ago. Looking west from any of the crater overlooks, you can see layers of Moenkopi near the crater's rim. Below it lies the gray, layered cliff band of 270-million-year-old Kaibab limestone

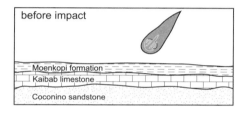

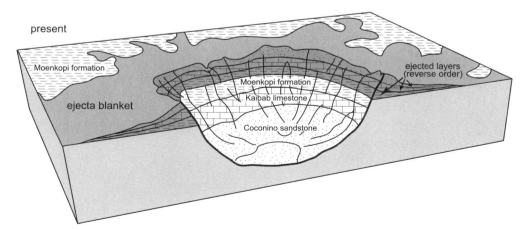

The impact of the 300,000-ton iron meteorite on the plains of northern Arizona 50,000 years ago threw rock debris onto the crater rim in the reverse order of the bedrock layering.

(discussed in vignette 7). Below the Kaibab, a blonder stone makes up the cliff to the north and east and continues to the crater's floor. This is Coconino sandstone, 275-million-year-old rock formed from ancient sand dunes.

Now look to the right from any of the visitor center overlooks toward the rim. Like Barringer, you'll see that the debris piled on the rim is separated into layers—in the exact *opposite* sequence of the local bedrock. Above the highest clearly defined red Moenkopi bedrock lies what looks like red soil: pulverized Moenkopi formation. Above this, a layer of gray Kaibab debris contains many large blocks. You can examine this layer closely; the vista points at the visitor center are built on it. Debris dominated by blond Coconino sandstone lies at the top of the crater's rim and on its outside flanks. Barringer noted that this sequence is exactly what one would expect from a projectile impact from above, blasting first through the Moenkopi, then the Kaibab, and finally the Coconino. And it is not what you would predict from a steam explosion from below, which would mix rock from all the formations in a chaotic jumble.

Barringer's arguments swayed a handful of professional geologists, but many remained skeptical. If this indeed was a meteor crater, where was the meteorite? After World War II, a meteorite collector named Harvey Nininger helped shed light on this conundrum. Nininger and his son had discovered spherical metal droplets in the soil around the crater. Nininger concluded that these droplets, now called spherules, showed that the meteor had vaporized upon impact. The spherules were vaporized material that had congealed in midair.

During the 1960s, Eugene Shoemaker and colleagues finally put an end to geologists' skepticism. They discovered a unique type of quartz at the crater, a mineral called coesite that forms only at extreme temperatures above about 1,300 degrees Fahrenheit and at pressures 20,000 times greater than those on the earth's surface. A titanic concussion must have created this mineral because volcanic eruptions don't raise pressure anywhere near that high. In a 1963 paper, Shoemaker presented a detailed study of Meteor Crater and compared it, feature by feature, to his equally detailed mapping of a crater made in the Nevada desert by an atomic bomb test. The paper swept away final doubts about Meteor Crater's origin. That year, Meteor Crater became the very first undisputed meteorite impact crater recognized on our planet.

Today, we can paint a picture of Meteor Crater's creation in some detail. At the moment of impact, rocks in the target area were fused by intense compression, squeezing Coconino quartz grains into coesite. A shock wave traveled back into the meteor, vaporizing all but a thin shell on its trailing edge where temperatures and pressures had not reached

From the crater rim, you can see buckled bedrock layers and the jumble of debris, some blocks of which are larger than cars.

such extremes. What was left of the meteor was blown back out of the crater by the expanding shock wave, along with 175 million tons of rock from the impact site. Shock waves propagate, or spread, spherically away from the collision that triggered them. The meteor's downward-propagating shock front continued to fuse, vaporize, and excavate rock to form the crater. Excavated debris known as the ejecta blanket quickly covered the surrounding area. Having been ejected first, the red Moenkopi fragments were the first to rain down, quickly followed by blankets of gray Kaibab, then blond Coconino fragments, some larger than a house. At the crater, trails to the overlooks wind around some of these blocks; you can get a sense of just how big they are and how much force it took to fling them out of the ground.

Inevitably, the shock wave waned as it penetrated deeper into the earth, lacking the punch to eject debris but maintaining the potency to fracture bedrock. About 1,300 feet below the surface, it ran out of energy to do even this, leaving the deeper rock layers undisturbed. Meanwhile, aided by the outward-propagating shock wave, the compressed rocks rebounded; layers along one-third of the rim even overturned. This accounts for the approximately 100-foot elevation of the crater's rim.

Geology Underfoot in Northern Arizona

Impact research continued. With advances in the understanding of high-velocity impact craters, the long-simmering lunar crater controversy was settled. Data collected by Apollo astronauts confirmed that impacts caused the thousands of craters we see up there, and it was immediately clear that craters on Mars and Mercury had been created in the same way. All three of these near-earth neighbors are pockmarked with thousands of craters ranging from a relatively diminutive 0.5 to 1 mile in diameter, the size of Meteor Crater, to over 700 miles across.

Scientists quickly realized that earth couldn't possibly have escaped bombardment similar to that endured by our neighbors. Earth's atmosphere protects us from most small projectiles, burning them up or slowing them. But our atmosphere clearly can't always protect us from the likes of the 130-foot blob of iron that gouged out Meteor Crater. Some scientists estimate that, on average, a meteor the size of the one that formed Meteor Crater visits the earth once every 1,600 years. Given the earth's 4,600-million-year age, something like 2.8 million Meteor Craters must have dimpled our planet's surface. For smaller bodies, it's more frequent: an estimated one thousand meteors weighing a quarter of a pound or more hit the earth's atmosphere every year.

In comparison, collisions with really large asteroids and comets, collectively known as bolides, are rare. Scientists estimate that a bolide big enough to gouge out a crater ten times larger than Meteor Crater slams into the earth an average of every 400,000 years. Collision with a bolide 6 miles in diameter, which can blast out a crater 100 miles across, occurs, on average, once every 100 million years, so even an event this infrequent has taken place some 46 times during earth's history. With so many collisions, why doesn't earth's surface show the bumps and bruises of the moon and Mars? Unlike those bodies, but like most state highway departments, the earth is continually resurfacing. Plate tectonics subducts pockmarked oceanic crust, mountain building crumples the earth's surface, erosion erases some craters, and deposition buries others thousands of feet below the surface, where they become impossible to detect.

As soon as they were learned, the lessons of Meteor Crater were applied around the globe. To date, about 160 impact craters have been confirmed, and the number is growing. Researchers also discovered an important link between oil and mineral deposits and impacting. A massive impact at Sudbury, Ontario, produced ores of nickel, platinum, and copper valued at over $100 billion. The Vredefort crater in South Africa, one of the largest, oldest craters so far discovered, has produced billions of dollars worth of gold. Impact fracturing of bedrock has also turned a number of craters into oil traps. Mexico's famous Chicxulub (CHEECH-zhoo-loob) crater, located off the Yucatan Peninsula, is a prolific producer of petroleum. Some geologists claim this massive impact 65 million years ago heated the organic-rich shales under the Caribbean Sea and drove the oil from them. If true, this makes the impact responsible for virtually all of Mexico's oil reserves—the country's economic lifeblood.

Chicxulub is also the focus of the most dramatic impact research since the confirmation of Meteor Crater's origin. In 1980, a group of scientists led by the father-son team of Luis and Walter Alvarez documented a high concentration of the element iridium in a curious clay layer they discovered in Gubbio, Italy. This layer was dated at 65 million years old—exactly the boundary between Mesozoic and Cenozoic time, also known as the K-T boundary (for Cretaceous and Tertiary, the periods straddling it). The Alvarez team claimed that this spike of iridium, which is much more abundant in asteroids than in earth rocks, could only have been produced by a massive bolide impact. They went on to claim that this impact caused a mass extinction known to have occurred at the same time, with which geologists in fact define the end of Cretaceous time. Most famously, this extinction event killed off the dinosaurs, but it also eradicated an astounding 70 percent of creatures living on the planet at that time.

Many geologists, biologists, and paleontologists have tried to explain the reasons for this mass dying. Not all of them welcomed the Alvarez hypothesis that a bolt from the blue was the cause. There was no "smoking gun," they objected—no known crater large and recent enough to account for such an impact. The discovery of the Chicxulub crater was a turning point. Because the crater had been filled in by 65 million years of sedimentation, its detection was much more difficult than at Meteor Crater. The information gained at Meteor Crater greatly facilitated its discovery.

With time, more and more earth scientists have come to accept a bolide impact as the cause of the K-T extinctions. But how could an impact lead to the cessation of 70 percent of life on earth? Recall that the collision that created the 0.75-mile-diameter Meteor Crater released energy equivalent to 2.5 million tons of TNT. The energy released by the Chicxulub impact, which left a crater over 100 miles across, far exceeded the combined energy of all of earth's nuclear arsenals, drastically amplifying the effects on the environment. Besides the direct killing of all organisms near ground zero, debris thrown into the air blotted out the sun for months, causing photosynthetic plants—the basis of the food chain on land and in the sea—to die. Their deaths had a ripple effect throughout animal communities.

The lack of sun drastically cooled the planet for a couple of years; many areas may not have warmed above freezing the entire time. In addition, the impact vaporized billions of tons of rock, most of which was limestone, which contains abundant carbon and some sulfur. This injected huge amounts of carbon dioxide and sulfur dioxide into the atmosphere. The sulfur dioxide, which reflects sunlight back into space, further cooled the planet for several years, until rain scrubbed it out of the atmosphere. In contrast, the carbon dioxide warmed the planet, just like human emissions today. The carbon dioxide lingered far longer, so that the world emerged from its deep freeze into a hothouse environment. As if these conditions weren't difficult enough, both gases, combined with water in the atmosphere, produce acid rain, which modern human experience has shown is toxic to many organisms.

Recognition of the effects of the Chicxulub collision by most scientists led naturally to the question of whether a meteor impact caused any other of the five major extinctions on earth. While there is nothing resembling scientific consensus, at least some evidence exists for impacts triggering all five, causing geologists and biologists to regard the evolution of the planet, and life itself, through a whole new lens. In particular, evidence has been growing for an impact offshore from western Australia or Antarctica as the cause for the massive P-T (Permian-Triassic) event 251 million years ago, the biggest extinction ever, which extinguished an incredible 90 percent of all life-forms—enough to threaten the existence of life on earth.

Why the dinosaurs, who dominated the globe for over 100 million years, suddenly died out has been an enduring enigma. Many scientists

A 1,400-pound fragment of the iron meteorite that gouged out Meteor Crater on display in the Meteor Crater museum

have claimed they simply became maladapted to their environment. While maladaptation does occur, this was a numerous, diverse, and successful group of organisms that all died out simultaneously. More and more, paleontologists and evolutionary biologists are concluding that the dinosaurs' demise was due not to evolutionary necessity, but rather, a cruel twist of extraterrestrial fate. With the dominant dinosaurs out of the way, our mammal ancestors—small beasts whose scavenging habits better equipped them to survive the impact—were free to experiment with evolutionarily new body plans that allowed them to radiate into open ecological niches. One of those experiments led through a series of steps to us. It is possible that we are the product of an evolutionary happenstance. The role random chance plays in evolution has always been acknowledged at the microscopic level of individual genetic mutations. Now, thanks in large part to advances in impact science that began here at Meteor Crater, we have begun to appreciate it on a macroscopic scale, at the level of the survival of a species or an entire ecological community.

Created in the blink of an eye, northern Arizona's Meteor Crater has had long-lasting repercussions. This hole in the ground has deeply impacted our understanding of our celestial neighbors, our planet, and our own evolution.

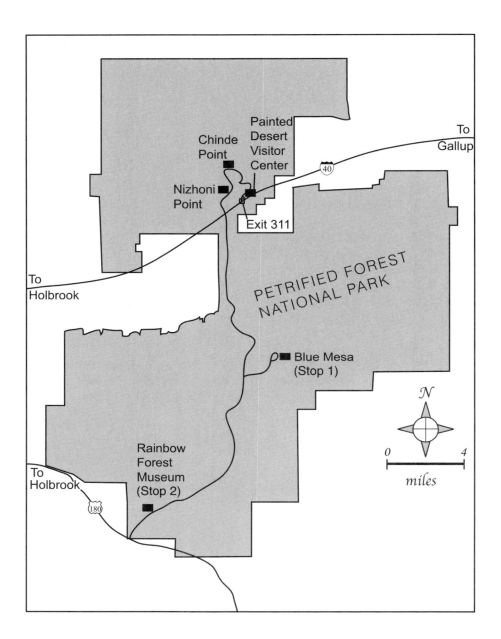

◆ GETTING THERE

Petrified Forest National Park is 91 miles east of Flagstaff. The park straddles I-40; take exit 311 to the northern entrance; there is also a southern entrance from U.S. 180. The park is closed from sunset to sunrise to combat the theft of petrified wood. It is a federal crime to remove objects from the park, and rangers have the right to search you and your vehicle for contraband wood. As you enter, remember to declare any petrified wood you have purchased outside the park. Stop 1 is a hike on the Blue Mesa Trail. From the north entrance, take the main park road 15.8 miles south and turn east onto the marked Blue Mesa road

15

Triassic Time Capsule
PETRIFIED FOREST NATIONAL PARK

The history of our planet is like a tattered scroll, and each rock or fossil is a scrap. If you could but locate and reassemble all the scroll's fragments, earth's 4.6-billion-year sweep of history would be revealed. But plate tectonics, weathering, and erosion continuously resurface the earth, removing whole chapters of the story in the rock and effectively shredding other parts into fragments too small to reassemble with confidence. These same geologic processes recycle other rock units into new rocks—secondhand pieces of parchment upon which more recent chapters are written. Geologists and paleontologists gain their greatest insights about earth history from the few rock units that have serendipitously survived the ravages of time with their stories nearly intact.

No rock unit in the world better records the events and cast of characters of middle Triassic time 234 to 209 million years ago than the Chinle (pronounced CHIN-lee) formation here in Petrified Forest National Park. This remarkable formation has yielded the remains of over two hundred species of plants and sixty species of animals. Many, like the park's famous trees, are exquisitely preserved, affording us intimate glimpses into daily interactions between Triassic plants and animals and their environment. Using subtle sedimentary clues and these incredible fossil finds, geologists and paleontologists have been able to reconstruct not just a landscape, but an entire ecosystem.

Stop 1, at Blue Mesa, allows us to piece together what the middle Triassic landscape looked like. Driving there from the north park entrance, you will pass nearly continuous outcrops of Chinle formation—for the most part, multicolored, tepee-shaped badlands nearly devoid of vegetation. It is spectacular scenery, but none of the petrified wood for which

---

for 2.8 miles. A parking lot and sun shelter mark the beginning of the Blue Mesa Trail. The 1-mile loop trail begins with a steep descent off Blue Mesa, then becomes more gentle. Allow about one hour to complete the loop. The features described in this vignette can be seen within the first 200 yards of the loop. Stop 2 is the Rainbow Forest Museum, 1 mile north of the south park entrance station. The gentle, 0.4-mile Giant Logs Trail is accessible through the museum.

Tepee-shaped, multicolored badlands lie beneath the harder, mesa-capping Sonsela member. Rapid erosion of the badlands has undercut the Sonsela in places, leaving behind rock towers in the left middle distance.

the park is named will be in evidence until you reach Blue Mesa. Flat-topped rather than tepee-shaped, Blue Mesa hosts the park's northernmost samples of petrified wood. The fact that the mesa is capped by a distinctive layer within the Chinle, the Sonsela member, explains both its shape and the presence of the wood.

From the Blue Mesa trailhead overlook, you can see that the Chinle's more typical pastel-colored badlands li e a short distance below the mesa rim, underneath the Sonsela member. The first rocks you encounter as you hike down the trail consist of Sonsela gravels with well-rounded particles. A few chunks of petrified wood are visible embedded in these gravels. In fact, the vast majority of the park's petrified wood is embedded in the Sonsela member for reasons we shall soon explore.

It takes a high-velocity current to move pebbles like these, and only a few sedimentary environments have enough energy to do the job. These include rivers, glaciers, landslides, and submarine canyons. The very well-rounded nature of these pebbles and the fact that they are fairly uniform in size helps us narrow down the list. Pebbles become round by banging against one another during long-distance transport. Ice in

Gravel of the Sonsela member lines the beginning of the Blue Mesa Trail. Petrified wood, like the piece seen here, is concentrated in this rock unit.

glaciers and mud in landslides tend to cushion pebbles from collisions with each other, and neither mechanism transports material terribly far. So pebbles transported in these ways are usually angular, not rounded. Furthermore, glaciers and landslides dump particles of many different sizes together, resulting in sedimentary deposits less uniform in particle size than this. Therefore, the Sonsela must have been deposited by a river or in a submarine canyon. The abundance of terrestrial fossils, including petrified wood, and the absence of marine fossils in the Sonsela indicate that the member was deposited by a river, not in a submarine canyon.

Continuing down the trail, you quickly leave the pebbles of the Sonsela member and encounter outcrops of typical Chinle badlands. The particles of this extremely soft material consist of mud; this is a mudstone. In contrast to gravel, mud is usually laid down in quiet waters such as lakes, swamps, deep ocean, and river floodplains. In places along the trail, thin layers of sand or gravel reminiscent of the Sonsela above are embedded in the mudstone. These coarser layers suggest the Chinle mudstone was not deposited in a lake or on the ocean floor, because these larger particles would have settled out along the ancient shoreline. Rather, these clues show that the Chinle mudstone was deposited either in swamps or along murky river floodplains.

Numerous plant and animal fossils have been discovered in the Chinle mudstones, including ferns, mosses, freshwater clams, and the fronds of palmlike cycad trees. You can see wonderful examples of these fossils in the museum at stop 2. These fossil species are ones you would expect in a floodplain or swamp environment, confirming our interpretation.

Our observations from the Sonsela and the underlying badlands lead to an illuminating portrait of the area during middle Triassic time. The plant fossils are of tropical varieties, indicating that the region lay near the equator and consisted of a vast, muggy lowland covered by tropical swamps and lakes and the floodplains of sluggish rivers. These had their sources in highland areas in Texas, New Mexico, and southern Arizona, and they flowed north from here into Utah and on to Nevada, where they finally met the sea. Massive quantities of mud accumulated on the lowlands—thick tropical river and swamp sediments now forming the Chinle badlands. A series of floods carried sand and pebbles from the higher regions into the swampy lowlands, depositing the Sonsela member. These floods uprooted trees and rafted them into the lowlands, depositing them with the gravels. Eventually they petrified. Petrified wood here, therefore, is found in conjunction with Sonsela flood deposits.

Blue Mesa contains the northernmost petrified wood deposits in the park because it contains the northernmost outcrops of the Sonsela member. The dearth of wood further north is due to the fact that the Chinle formation tilts gently down to the northeast in this area, meaning that, as you travel from north to south along the park road, you encounter progressively older layers. The uppermost Chinle beds at the park's northern end date to about 209 million years ago. In contrast, Blue Mesa consists of rock closer to 227 million years old. The younger Chinle rocks to the north contain few flood-derived gravels due to a general drying trend later in Triassic time.

Why is Blue Mesa flat-topped whereas the badlands form triangular hills? Like all of the park's mesas, a relatively resistant rock layer caps Blue Mesa—in this case, the Sonsela member. Erosion has completely removed softer mudstone layers that used to overlie the Sonsela here,

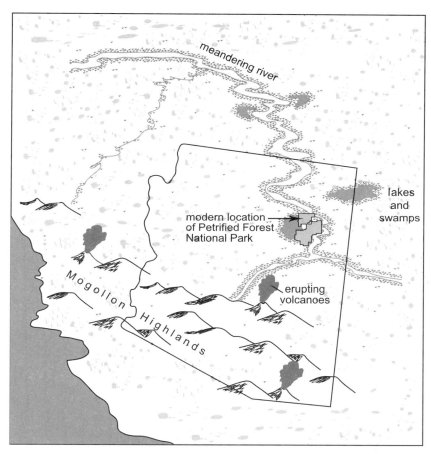
Reconstruction of Arizona's geography during middle Triassic time

forming a flat surface on top of the resistant unit. Were it not for the presence of hard layers like the Sonsela, there would be no badlands; they would have eroded long ago, along with their fossil treasures.

The Sonsela protected the soft mudstones around you here until several streams succeeded in breaching the Sonsela cap and began to quarry away the soft mudstones below, undercutting the cap and causing big slabs of it to tumble down the slope. The mesa edge thus retreats away from the wash in a process called scarp retreat, until the mesa disappears entirely (more on scarp retreat in vignette 4). What remains is a chain of triangular badland mounds like those that surround you. Soon these too will erode away, and the topography will flatten completely, with erosion pausing briefly on the next resistant layer in the sedimentary stack.

Perusing the badlands around you, you will immediately notice their scaly texture, resembling elephant hide, and their multicolored hues. Both these characteristics are the result of ash belched from a distant chain of volcanoes. The ash occasionally blanketed the swamps and lakes or was carried downstream by rivers and stirred into the mud.

Volcanic ash facilitates the formation of badlands due to its unusual chemical properties. Because it is inherently unstable and has small particles, ash is very susceptible to chemical weathering, which transforms it into a clay called bentonite. When it rains, bentonite absorbs massive quantities of water and swells to seven times its dry volume. The hot desert sun then desiccates the clay, eventually shrinking it to its original size and leaving behind the cracked, scaly texture you see here. These wild fluctuations make it extremely difficult for plants to take root. The mud's high salt content and the presence of toxic trace elements like selenium add to the difficulty.

Without vegetation to soften the blow of raindrops, the mudstone is incredibly vulnerable to torrential monsoon rains that rake the area every summer, literally washing these hills away. The badlands are eroding at the geologically shocking rate of 3 inches every ten years. In contrast,

Bentonite clay is responsible for the scaly appearance of these badland mudstones. The petrified wood and surrounding pebbles tumbled down from the Sonsela member above. Note the growth rings on the wood.

slopes of hard rock such as granite erode at a rate closer to 3 inches every ten millennia.

Volcanic ash also imbues the badlands with their striking swirls of color. When the colorless ash weathers, elements such as iron, manganese, magnesium, and sulfur leach from the ash into the groundwater, forming new, pigmented minerals. Even tiny traces of these elements produce the palette of yellow, brown, red, purple, and pink seen here. The dark gray mudstones derive their color from organic carbon left over from once abundant vegetation.

In addition to great views of the badlands, the Blue Mesa loop offers abundant petrified wood for you to examine in the trail's lower reaches. This wood and the rounded pebbles typically surrounding it tumbled down from the Sonsela bed capping the mesa. As scarp retreat whittles the mesa edge back over time, it liberates logs from the rock, and gravity takes over. Most of the logs here are small; more spectacular concentrations of wood await you at stop 2.

Begin your tour of stop 2 by walking the Giant Logs Trail behind the Rainbow Forest Museum. An interpretive brochure available at the museum orients you to eleven numbered stops along the way. Like the Blue Mesa Trail, this path winds through the Chinle's wood-rich Sonsela member. Everywhere you look, petrified logs litter the ground. The preservation of each tree's form as it slowly turned to stone is breathtaking. The wood's fibrous texture is still plainly visible, as are growth rings. Triassic trees did not produce annual growth rings; rather their rings represent more extended periods of wet and dry weather. Knots where branches met the trunk look identical to those on freshly felled trees. The logs are up to 150 feet long, yet none preserves an entire tree, indicating that, when alive, the trees probably soared 200 feet tall. Wood from three different species of cone-bearing trees has been found in the park, but by far the most abundant is from a conifer called *Araucarioxylon arizonicum*, a distant relative of modern Norfolk pines and monkey puzzle trees found today in the southern hemisphere.

Before the trees were deposited in the Chinle sediments, almost all their limbs and roots were broken off, and the bark was stripped. In addition, although a few stumps have been found rooted where the trees grew, the vast majority of fossil logs lie prone like driftwood—which is exactly what they are. The trees were felled by floods and washed downstream, battered and broken as they rushed along. They came to a stop in river-bend logjams, then were quickly buried by mud and sand deposited by waning floodwaters. This rapid burial helped preserve the trees; the entombing muds contained little oxygen to sustain the bacteria that drive decay of organic tissues.

Tiny details are preserved during petrification. Knots indicate where branches once protruded, growth rings are still visible, and the texture of the wood is lifelike.

Auricaria trees in Chile, modern relatives of Triassic trees —Wayne Ranney

Despite their rapid burial, most of the trees would still have slowly decomposed if not for the volcanic ash in the Chinle's river mud. As in the process that produced the rainbow of colors in the Chinle badlands, percolating groundwater dissolved the ash into its constituent chemical elements and distributed them through the sediment. In addition to iron and manganese, the ash contained the compound silica. When the silica-charged water permeated the buried logs, chemical reactions with organic material caused the silica to precipitate out.

Microscopic crystals of quartz called chert solidified in all of the pore spaces of the trees' cells, preserving the cell walls in a matrix of stone. This process is called permineralization. Permineralized logs tend to be brown in color. However, the chert deposited here had trace amounts of iron, manganese, and other elements in the ash. These trace elements pigmented the chert in myriad shades of red, yellow, blue, purple, and black. Such varicolored chert is known as agate, a beautiful and highly

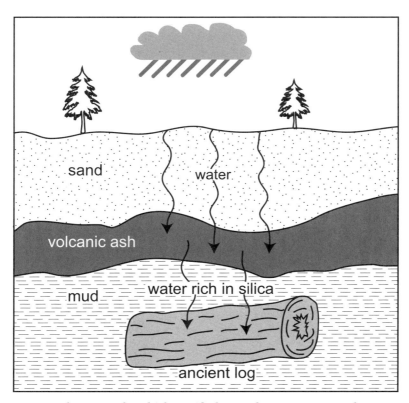

The process by which Petrified Forest logs were preserved

prized semiprecious stone. These colors make the wood of Petrified Forest National Park some of the most beautiful in the world. Cavities in the wood allowed space for larger crystals to grow, producing gem-quality specimens that greedy, insensitive visitors have carted off in alarming quantities.

Two numbered stops along the Giant Logs path are particularly worth a visit. Marker 8 is planted next to an especially large log dubbed "Old Faithful" by the wife of the park's first superintendent. Unlike most of the logs you see here, part of Old Faithful's root system is preserved, giving you a good picture of what the base of the tree looked like. What you see at marker 10 is subtle but even more amazing. Look closely at the wood here. Small grooves score the log's surface. These are actually petrified borings—excavated pathways made by bark beetles an astonishing 227 million years ago. No fossils of the beetles themselves have been found, but the borings of modern bark beetles are nearly identical, suggesting that the survival strategies of these creatures have changed little over hundreds of millions of years.

Elsewhere in the park, even smaller borings have been discovered in the stem of a fossilized tree fern; scientists believe the borings were excavated by leaf-eating mites. These borings are so well preserved that the mites' fossilized feces are still visible. Fossilized feces, known as coprolites, are surprisingly common in the geologic record. To learn who made a coprolite and what the creature ate, researchers study its shape and chemical composition. Much larger coprolites have also been found in the park. Because these larger ones weren't flattened when they hit the ground, they were likely deposited in water. Their shape and size and the presence of bone matter all indicate that they were expelled by carnivorous metoposaurs that lived in the Triassic swamps. These larger coprolites also appear to have borings preserved in them—evidence of the oldest known feces-eating insects.

If you could squeeze into a time machine and journey back to this same spot in late Triassic time, it would have a distinctive rain forest feel. In the higher areas, conifer trees draped in club mosses towered above the dim, damp forest floor. A lush understory of ferns, tree ferns, ginkgos, and cycads further filtered out the sun's powerful rays. Horsetails the size of telephone poles grew in low, swampy areas. Beautifully preserved specimens of several of these species are on display in the Rainbow Forest Museum.

There was, however, one huge distinction between this ancient rain forest and modern ones. In Triassic time, you would have scanned the scene from one horizon to the other without seeing a single flower. The flowering plants that dominate today's rain forests would not evolve for

During Triassic time, the Petrified Forest consisted of swamps verdant with cycads, ferns, and ginkgo trees. Better-drained areas supported a conifer forest, the remains of which are the famous petrified logs. Diverse, bizarre animals included the early dinosaur Coelophysis, *here being eaten by a phytosaur, and, clockwise from there, rhino-sized dicynodonts; procolophonids (left bottom corner); metoposaurs; aetosaurs (reminiscent of modern armadillos); and* Postosuchus. —Dona Abbott

another 100 million years. However, the Chinle formation contains fossils of what may be their direct ancestor. Known as *Sanmiguelia lewisii*, this 3-foot-tall plant has some flowerlike features, leading many paleobotonists to consider it a missing link between palms and the first flowering plants.

Despite the lack of flowers, petrified bee nests have been discovered in both Chinle soils and fossilized trees. These are the oldest bee nests ever found, pushing the group's evolutionary appearance back a whopping 100 million years earlier than scientists had previously believed. The similarity in structures between modern and fossilized nests suggests that bee behavior has changed very little in almost 230 million years. Even that long ago, bees appear to have been social insects, and the fossils indicate that they survived by collecting pollen. Many scientists had speculated that pollinating insects and flowering plants appeared on the scene together, but evidence from the Chinle nests suggests that early bees subsisted on pollen from conifers and cycads.

Sanmiguelia lewisii, *a forerunner of flowering plants* —Dona Abbott

The Triassic was a time of transition for terrestrial animals. The period was ushered in by the granddaddy of all extinction events, which killed an astounding 90 percent of earth's species 251 million years ago. As witnessed by the sixty species of animal fossils that have been discovered in the park, life rebounded from that catastrophe with an almost entirely new cast of characters. Crayfish, turtles, horseshoe crabs, and numerous species of fish thrived in the ponds and rivers in the Petrified Forest lowlands. Clams were especially abundant, with some Chinle layers up to 2 feet thick and 1,000 feet long composed of nothing but thousands of clams packed cheek to jowl.

Freshwater sharks and some of the world's first crocodiles prowled the rivers and swamps. They were dwarfed by big lowland predators, including flat-headed metoposaurs up to 10 feet long and half a ton in weight—some of the largest amphibians ever to inhabit the earth. The nastiest predator of all was *Postosuchus*. This agile reptile resembled a 13-foot-long alligator, but sported much longer legs, rapier-sharp claws, and an oversized head equipped with rows of sharp teeth. It dined on aeteosaurs, armor-plated, armadillo-like reptiles that inhabited the swampy lowlands. You can examine the spade-shaped head of a metoposaur and a full aeteosaur skeleton in the museum.

Postosuchus, *the largest, nastiest predator in the Petrified Forest* —Dona Abbott

The Triassic also witnessed the dawn of the dinosaurs. In 1982, the park's first dinosaur fossil was discovered in the northern badlands. It was a small, ravenous meat-eater called *Coelophysis*, 7 to 8 feet long with razor-sharp claws on both hands and feet. A swift predator, *Coelophysis* probably hunted in packs, chasing down small amphibians and the world's first lizards, also recently evolved.

Coelophysis and other early dinosaurs played a fairly minor role in the Petrified Forest ecosystem, but that would change dramatically when another major mass extinction hit at the close of Triassic time. Many of the park's plant and animal species died at that time, but based on tracks found throughout northern Arizona, *Coelophysis* or a near relative not only survived, it thrived in this new world no longer dominated by the likes of *Postosuchus*. Lacking competition, this species quickly ascended

One of the world's first dinosaurs, Coelophysis —Dona Abbott

to the top of the terrestrial food chain, a position that *Coelophysis* and its larger and more fearsome descendants (vignette 8) would not relinquish for 135 million years.

The level of detail preserved in the Petrified Forest trees is so awe inspiring that some visitors have a hard time believing at first sight that they are truly made of rock. An even closer look at the park's petrified trees and myriad other fossils reveals ever finer details about what life was like for the plants and animals living under the canopy of this stately forest.

Petrified Forest's exquisitely preserved fossil record is filled with amazing glimpses of Triassic ecological interactions. From the Chinle formation's fossils, geologists and paleontologists have woven a vivid tapestry of Triassic land life. These glimpses demonstrate not only how much has changed on earth since Triassic time, but perhaps more surprisingly, how much, like the bark beetles, has remained the same. A trip to the Petrified Forest is like opening a Triassic time capsule, unlocking a window on an ancient world at once exotic and thoroughly familiar.

16
Continent Under Construction
THE ROCKS OF PRESCOTT

The rocks of northern Arizona tell many fascinating tales—stories of dinosaurs and long-extinct trees, of meteorites and volcanic eruptions. But the rocks around Prescott tell an older story: the birth and evolution of a continent. The world was utterly different then. No animals existed; the air was unbreathable poison. Still, one crucial characteristic operated the same way it does today: plate tectonics.

The theory of plate tectonics revolutionized geology in the 1960s. It asserts that the earth's surface is not as static as common sense and everyday experience would have us suppose. Rather, the surface consists of twelve major and many smaller rigid plates that move across the face of the globe, wafted by currents in the solid but plastically flowing asthenosphere—that portion of the earth's mantle approximately 50 to 150 miles beneath the surface. The plates can move toward and away from each other or slide laterally past each other. The crust that composes them comes in two varieties, oceanic and continental. As their names imply, dense oceanic crust forms the deep ocean basins, and more buoyant continental crust makes up the world's continents. Oceanic crust is produced where two plates pull apart and plumes of molten lava from the asthenosphere spew forth to fill the void. The resulting dense black rock is known as basalt.

When plates of this oceanic crust meet—at a boundary usually marked by a trench and known as a subduction zone—the older, colder, and thus denser of the two dives hundreds of miles beneath the other, deep into the bowels of the earth. There it heats up and partially melts. Andesitic magma generated by the melt (more in vignette 11) rises to the surface and erupts from a chain of towering island volcanoes that, due to the geometry of spheres, on a map trace an arc following the plate boundary. Geologists refer to them as volcanic arcs. The erupting lava possesses more of the chemical compound silica than basalt does, making it less dense and therefore more buoyant than oceanic crust. This material rides above the ocean basins, floating on the hot, plastic asthenosphere, where it composes the continents and is therefore called continental crust.

Rafting around the globe on their respective plates, volcanic island arcs sometimes find themselves on a collision course. As oceanic crust

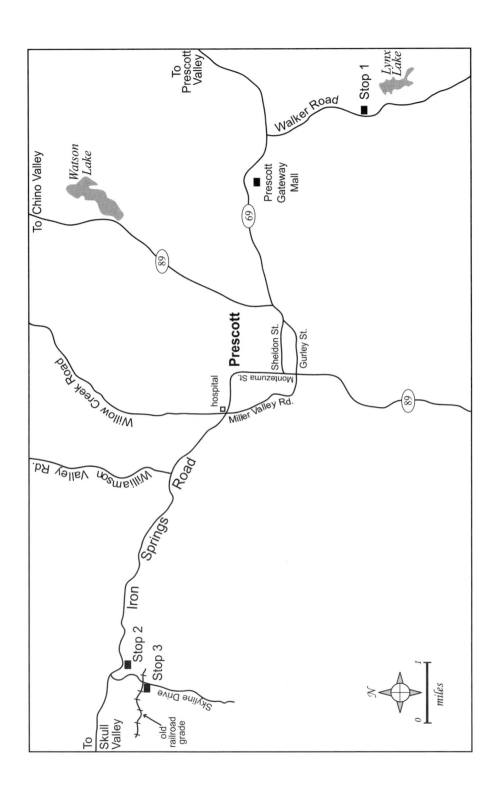

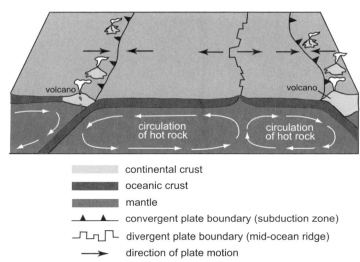

The movement of earth's plates influences the production of oceanic and continental crust.

attached to one arc subducts beneath the other, it eventually drags the islands into the oceanic trench marking the subduction zone. The arc's continental crust, however, is too buoyant to dive hundreds of miles into the earth's interior. Instead, it rides up over the edge of the other, squeezing both arcs' rocks like the metal of two car hoods in a head-on collision, and thrusting up a mighty mountain range. The resulting swath of land is a microcontinent. As time goes on, microcontinents collide and amalgamate into continents.

In this way, North America began to assemble a staggering 4,000 million years ago in the Hudson Bay area. By 1,800 million years ago, it had added

GETTING THERE

Stop 1 is the Highlands Center for Natural History. From downtown Prescott, travel southeast on Arizona 69 for 4.5 miles to Walker Road, just east of the Prescott Gateway Mall. Turn right (south) for 2 miles to a parking lot on the left. Walk down the ramp to the covered amphitheater, where the center's trail system begins. In this vignette, you will hike a fairly gentle 1.3 miles along these trails. To reach stop 2, go west, back toward town, on Arizona 69. At a clearly marked fork, go right on Sheldon Street to a T-junction. Turn right onto Montezuma Street, which becomes Iron Springs Road. After 1.5 miles, you reach the hospital and a major intersection with Miller Valley Road. From this intersection, continue straight for 4.9 miles. Pull into a large gravel pullout on the left to examine the roadcut across the highway. For stop 3, continue another 0.3 mile northwest on Iron Springs Road and turn left onto Skyline Drive. Proceed south for 0.7 mile to a gravel road on the right, and park in front of the gate.

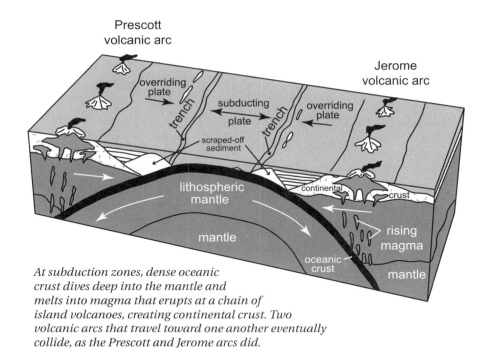

At subduction zones, dense oceanic crust dives deep into the mantle and melts into magma that erupts at a chain of island volcanoes, creating continental crust. Two volcanic arcs that travel toward one another eventually collide, as the Prescott and Jerome arcs did.

Mount Augustine, which forms its own small island in Alaska's Aleutian chain, is a classic subduction-zone volcano.

most of Canada, the Great Lakes area, and a long, westward appendage including present-day Wyoming, Montana, Utah, and southern California. The scene was thus set to plaster the next chunk of material onto the continent, a swath that would include northern Arizona.

As the curtain rose on this story 1,800 million years ago, the rocks that today comprise northern Arizona were just being forged in a series of subduction zones. Amidst a tangle of mountainous island chains reminiscent of today's Southeast Asian archipelagos, one subduction zone began to build a chain of volcanic islands in the Prescott area, which we will call the Prescott arc. A similar but separate chain, which we will call the Jerome arc, was simultaneously developing several hundred miles away. Subduction of the intervening oceanic crust caused the two arcs to viciously collide some 1,700 million years ago, forming a microcontinent (see top figure on page 243). Almost immediately thereafter, this microcontinent was welded to the edge of early North America in yet another collision.

At stop 1, get on the 0.3-mile Stretch Pebble Loop Trail (#443) to begin exploring evidence of these continent-building collisions. Twenty yards into the hike, on the left across from interpretive marker 1, is a low, brown outcrop displaying pronounced, vertically oriented planes—what geologists call a fabric. This is a type of metamorphic rock called phyllite, a close cousin to schist, but with smaller, harder-to-see mica crystals. Like all metamorphosed materials, phyllite forms when a rock subjected to elevated temperature and pressure is transformed physically and/or chemically without melting in the process. These conditions often result in alignment of micas, with their characteristic silky sheen. It is this alignment of micas that created the vertical fabric of this rock. Metamorphic rocks offer clues to two separate events in their history: the conditions under which the parent material formed and how the rock was transformed.

Rocks like phyllite are common around the Prescott area. Microscopic examination of the collection of minerals it hosts suggest that its parent material was volcanic ash—our first confirmation of the existence of the Prescott volcanic arc. With their high silica content, lavas erupting from island arcs are sticky, efficiently trapping the gases that rise with the magma. Gas pressure builds in the magma chamber until the volcano violently erupts, spewing its lava high in the air, where it forms volcanic ash (vignette 10).

The fact that this rock is phyllite tells us more about how the ash was metamorphosed. Phyllite forms under differential pressure—pressure that is not equal from all directions. Such pressure characterizes subduction and collision zones, where the rocks are squeezed most in the direction the plates are moving. The micas that form in response all grow in the same direction, with their elongated crystals running perpendicular to the direction of squeezing. This parallel mineral alignment is called foliation, a reference to its resemblance to the pages of a book.

Watch for another outcrop of phyllite further along the trail, halfway between markers 1 and 2. At the junction with trail #305, about 600 yards from the trailhead, continue straight through on trail #443. Immediately beyond, the trail branches in a short loop around a low hill; either direction brings you to an outcrop for which the Stretch Pebble Loop Trail

The Prescott and Jerome volcanic arcs and early North America 1,800 million years ago (below) *resembled modern Southeast Asia* (next page), *where the Australian continent is bearing down on a complex of volcanic arcs.*

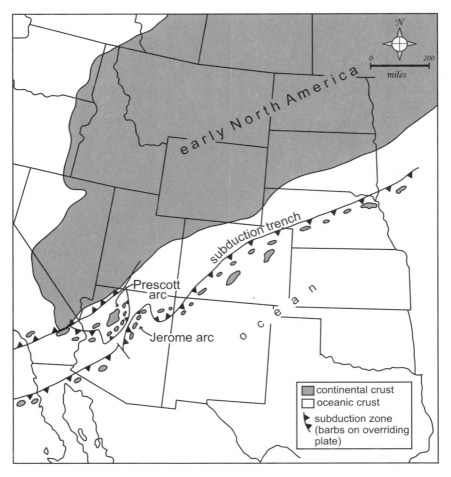

is named, at interpretive marker 7. This beautiful rock was originally a pebble conglomerate, but metamorphism stretched the previously round pebbles into white, red, and gray streaks. Just as in the phyllite, the stretched pebbles form a vertical foliation.

This outcrop provides clues to conditions at the time of metamorphism. Note that its vertically inclined beds of blocky, resistant conglomerate alternate with low, dirty sections of gray or brown phyllite. For example, the low portion of the outcrop to the right of the #7 sign consists of brown phyllite on its left edge, a thick conglomerate bed in the center, a section of metamorphosed sandstone hosting just a few pebbles to the right, and a thin pebble-rich layer to the far right. These are original layers of, from left to right, ash, conglomerate, sandstone, and conglomerate that accumulated prior to metamorphism.

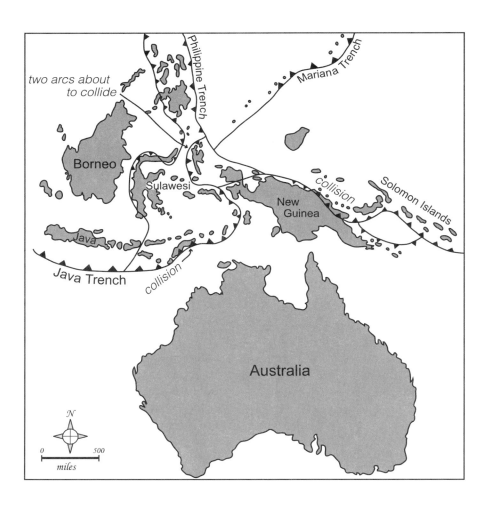

The stretch-pebble conglomerate's foliation, as indicated by the direction in which the pebbles were stretched, is parallel to these original sedimentary layers. The rock layers were subjected to such extreme folding that originally horizontal layers have in most places been tipped vertical. In such extremely tight folds, known as isoclinal folds, the hinge areas (each U-shaped bend) often erode away or are buried beneath the land surface, leaving a stack of vertically bedded and foliated rocks like those you see here.

Because these layers have endured the ravages of isoclinal folding, it is hard to know if the conglomerate sections in front of you are separate beds, or if you are looking at just one or two layers that are crumpled like an accordion. In either case, the primary source of such extreme squeezing is the collision between two volcanic arcs or continents. Both the bedding layers and the foliation here are aligned in a north-south direction, indicating that the squeezing took place in an east-west direction.

In the metamorphosed conglomerate to the left of the quarter, the stretched pebbles are barely recognizable. Metamorphosed sandstone lies to the right of the quarter, and a thinner layer of stretch-pebble conglomerate forms the outcrop's right edge. Foliation in the stretch-pebble section is parallel to the original bedding layers, a sure sign these rocks have been isoclinally folded.

The presence of a major fault zone between here and Jerome, the Shylock fault zone, tells us the culprit was the Jerome arc to the east.

The types of pebbles in the conglomerate reveal more about pre-metamorphic conditions. The red pebbles are chert, a sedimentary rock usually deposited in an ocean. The translucent gray pebbles are quartzite, a metamorphosed sandstone, suggesting the region experienced an even earlier metamorphic episode. And the abundant dull white to light

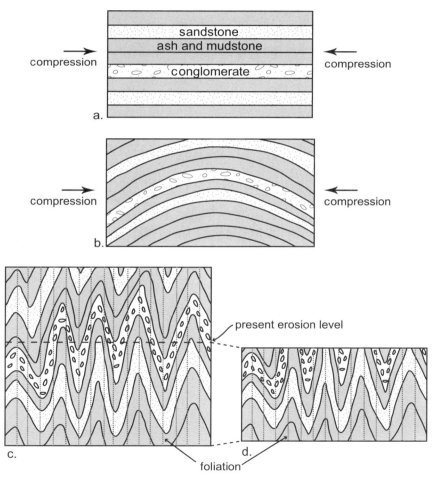

As a stack of sedimentary rocks is squeezed in a subduction or collision zone, the layers buckle into tight folds and the rocks begin to metamorphose, with the minerals aligning themselves vertically, perpendicular to the compression direction. If the squeezing is intense enough, the folds become very tight, conglomerate pebbles stretch, and the bedding layers become parallel with the foliation.

tan pebbles appear to be hardened volcanic ash, or tuff, corroborating the area's origins in a volcanic arc.

Complete the Stretch Pebble Trail's loop and return to its junction with the Homestead Trail (#305). Turn right (north) for about 250 yards to the junction with the Highlands Trail (#442) and turn right up a small hill. Within the first 30 yards the ground becomes littered with fist-sized chunks of maroon to dark gray rock. These chunks become larger and more abundant as you continue up the hill. If you pick one up, notice how surprisingly heavy it is. This is due to its extremely high iron content, typically around 30 percent. Many of these rocks lack obvious layering, but a significant number display alternating red and gray laminations only a few millimeters thick. In some of the rocks, the laminations are tightly folded.

These rocks are known as iron formations, and the ones with laminations are a special type called banded iron formations, or BIFs. They were deposited in quiet ocean waters, far from any disturbances that could erase their delicate bands. The iron is bound up in the gray layers, which consist of dense minerals such as magnetite and hematite, whereas the red layers are jasper, a type of chert whose color is derived from tiny amounts of rusted iron.

BIFs are extraordinary. Their very existence testifies to a radically different atmosphere. Because of iron's tendency to chemically bond with oxygen, BIFs don't form in today's oxygen-rich world. Iron atoms liberated on land by weathering are quickly bound up in new oxygen-bearing compounds, preventing their transport to the sea. Today's ocean waters

A banded iron formation. The dark layers are red jasper; the lighter bands are hematite, an iron ore.

possess only tiny amounts of dissolved iron—far from enough for BIFs to form.

BIFs were deposited in great abundance between 2,500 and 1,800 million years ago, but then their formation abruptly stopped. They enjoyed a small renaissance between 850 and 650 million years ago, but no BIFs younger than that have been discovered. Geologists believe this pattern of deposition actually traces the development of our atmosphere through the ages. The early abundance of BIFs indicates that earth's original atmosphere was almost wholly lacking in free oxygen. This conclusion sits well with planetary scientists, who don't detect oxygen in the atmospheres of any other planets in the solar system. Instead, they say, the early atmosphere consisted of nitrogen, carbon dioxide, water vapor, and methane. How, then, did earth's atmosphere evolve into the oxygen-rich mixture we breathe today? And how did the BIFs form? The surprising answers point to life itself.

Bacterial life evolved in the oceans sometime between 3,900 and 3,500 million years ago. One early group, the cyanobacteria, were the earth's first photosynthesizers and produced oxygen as a waste product. This oxygen was toxic to the cyanobacteria, so they could not have flourished had its concentration built up. Fortunately for them, dissolved iron existed in the oceans in abundance. This dissolved iron immediately bonded with the cyanobacteria's expelled oxygen to form the minerals hematite and magnetite, which quickly sank to the ocean floor, accumulating as the gray, iron-rich bands in the BIFs. As the bacterial colony continued to expand, it produced more oxygen than the local supply of dissolved iron could lock away, causing a toxic buildup of oxygen and mass mortality. After the colony's death, iron-poor red chert bands accumulated until bacteria repopulated the area, beginning the cycle anew. The bands you see in these rocks depict countless bacterial boom-and-bust cycles that are thousands of millions of years old.

By 1,800 million years ago, the cyanobacteria had produced so much oxygen that there was almost no iron left in the oceans to bond with; oxygen began to accumulate in the atmosphere instead. Over the long millennia, cyanobacteria had developed a tolerance for oxygen, so they kept right on photosynthesizing, but many less hardy species must have gone extinct at this time. The 1,800-million-year-old BIFs you see here record the last gasps of an earlier world, the way it was before the great oxygen crisis forever changed life on our planet.

East of the BIF hill, the Highlands Trail descends into a small ravine and crosses another low ridge before reaching the banks of Lynx Creek, where it turns right. Sixty yards after reaching the creek, the trail passes to the right of a small, finlike outcrop. This granite dike wriggled its way into the surrounding metamorphic rock around 1,700 million years ago, exploiting a plane of weakness in the foliation. An igneous rock with high silica content, granite solidifies deep underground where crystals have time to grow large. This granite dike is one of the many pipes through which the magma passed on its way to the surface to erupt volcanic ash—the parent material of the phyllite we saw earlier.

Immediately east of the granite fin, the rock is darker and flatter. This is metamorphic rock into which the granite intruded. The dark color and microscopic crystals indicate it's a scrap of an ancient basalt lava flow, and though metamorphism has obscured most of its original texture, enough remains to reveal more about the early Prescott landscape. Look in the center of the outcrop for an oval discoloration. This is the calling card of a special volcanic texture called a pillow, which forms only when basalt erupts underwater. As lava plunges into the ocean, cold seawater quickly cools its exterior, forming an oval crust into which still-molten lava continues to pour. Like an expanding Jiffy-Pop popcorn popper, the pillow puffs up until its rind cracks, diverting the lava stream to one side,

The curving white area just below the quarter defines the outline of a basalt pillow in this metamorphosed lava flow along Lynx Creek.

where it forms an adjacent pillow. Geologists have found other pillow basalts in this area, including a beautiful one displayed in Jerome State Historic Park (see vignette 17). Clearly, this was an island arc erupting in the middle of the ocean, like the Aleutians, rather than volcanoes on the edge of a continent, like the modern Cascade volcanoes of the Pacific Northwest.

A short distance beyond the pillow basalt, the trail exits Lynx Creek by climbing up a spring-fed tributary, which has conveniently polished an extensive outcrop of dark metamorphic rocks intruded by dikes of lighter granite. Many of the metamorphic rocks were originally basalt, and a few pillow structures are faintly preserved.

The trail climbs into a meadow. About 0.3 mile from the creek, you again intersect the Homestead Trail (#305). After turning right (north) onto it, you pass through another small meadow, then cross a tiny wash after about 100 yards. To your left, just a few feet up the wash, lies a low display of dark, green-tinted rock. This is another bed of metamorphosed basalt; its green tint is due to the presence of the mineral chlorite. This outcrop contains three pillow structures like the one you examined earlier. Look for other oval discolorations here. A further 250 yards brings you to the junction with the Stretch Pebble Loop Trail. Go left here to return to your car.

Remarkably, despite their long and tortured history, the rocks of Prescott yield ample evidence from which to paint a vivid picture of their origin. The rocks speak of a fiery birth 1,800 million years ago in a chain of explosive volcanoes straddling a subduction zone spanning hundreds of miles of open ocean. Lavas poured into the sea, puffing up stacks of basalt pillows. Plumes of ash billowed from the towering mountains, settling over the adjacent islands and the surrounding sea. During lulls in volcanic activity, erosion gnawed away at these volcanoes, depositing blankets of mud, sand, and pebbles. Generations of volcanoes were born, matured, and went extinct, only to be covered by the extrusions of their successors. More than 100 million years of volcanic activity produced a thick pile of buoyant continental crust.

Now we'll look for more evidence of the Prescott arc's collision with its similar neighbor, the Jerome arc. As you drive to stop 2, notice Prescott's abundant granite boulders. These are the solidified remains of a handful of the magma chambers that fed the Prescott arc's volcanoes 1,800 to 1,700 million years ago. Similar pods of granite underlie the volcanoes of active subduction zones, including the modern Cascades.

The roadcut at stop 2 consists of brown, vertically foliated phyllites and black metamorphosed basalt sliced by a strikingly pink granite dike. The dike, in turn, is cut by three faults angling down and left across the

Three reverse faults (marked by arrows) *cut this outcrop at stop 2. In reverse faults, the upper block moves up and over the lower block.*

outcrop's face. Along each of these faults, the left (upper) block has slid up and right. This signifies they are reverse faults, formed when the slab of rock overlying the fault is heaved up and over the slab below. Reverse faults form in response to tectonic squeezing, stacking slabs of rock one atop the other to accommodate compression. Thousands of such faults, big and small, rent both the Prescott and Jerome arcs during their collision.

Notice also the highly weathered orange granite on the extreme left side of the roadcut. This is the edge of a magma chamber that fed the volcanoes; the metamorphic rocks in front of you formed the wall of the chamber when the granite cooled and solidified over 1,700 million years ago. At that time, all these rocks lay at least 6 miles underground, but the force of the collision would heave the area up into a mighty mountain range. The range would then relentlessly erode, revealing these rocks on the earth's surface.

At stop 3, which lies along an abandoned railroad grade, work crews blasted a notch to make way for the train, revealing ancient rocks in the process. The same phyllites and metamorphosed basalts you have seen before are cut in many places by more granite dikes. As at stop 2, numerous reverse faults offset these dikes. The two most prominent reverse faults lie on the right, in front of the gate.

Look at the phyllite immediately to the right of the gate. A thin granite dike slashes through this outcrop, but instead of being straight, like the others you've seen, this one has been folded again and again, like the bellows of an accordion—just like the isoclinal folding we observed in the stretch-pebble conglomerate. The other granite dikes we've seen intruded the metamorphic rocks after the folding that deformed the conglomerate. This dike intruded before it. It was then folded with the rocks it intruded, resulting in these crazy zigzags. The older dike allows you to see just how tight those folds really are. Other phyllites we have seen were likely squeezed just as thoroughly, but without such a marker, we cannot see it.

As the Prescott and Jerome arcs collided, the inconceivable force metamorphosed the great stacks of volcanic and sedimentary rocks and squeezed them into tight isoclinal folds. The many reverse faults heaved a series of rock slabs one atop another, creating a craggy coastal mountain range. The two volcanic arcs were sutured together along the Shylock

The thin, light-colored granite dike has been isoclinally folded. Accompanying extreme stretching caused the dike to pinch out in places, but you can see tight folds immediately to the left of the quarter and farther away on either side.

fault, which is sprinkled with dismembered scraps of oceanic crust that once lay between them. The fault runs north-south along the east flank of the Bradshaw Mountains south of Prescott before breaking across the Chino Valley to nestle along the west flank of the Black Hills, the mountains west of Jerome.

The amalgamated Prescott-Jerome microcontinent barely had time to recover from this massive collision before being swept against early North America's encroaching mass. Many other small volcanic arcs existed in the area at that time, and the continent, which had been bearing down from the northwest, collided with one after another, like a windshield smashing bugs. These continuous collisions added a 300-mile-wide swath of land to the continent, stretching from northern Sonora, Mexico, across northern and central Arizona, to Colorado, Kansas, and Nebraska.

The geologists who first figured out plate tectonics envisioned an orderly assembly of continental building blocks; it wasn't until the 1980s that they realized messy chaos is the norm. A modern example of messy tectonics can be seen in Southeast Asia, in the tangle of small plates between the much larger Australian, Pacific, and Eurasian plates. Subduction zones hosting island arcs with volcanoes busily manufacturing continental crust separate many of the smaller plates. Impinging on this chaos, continental Australia sweeps northward. Some of the arcs are colliding directly with Australia, such as the island of Timor and the northern edge of New Guinea. Other arcs are heading for each other, just like the Jerome and Prescott arcs long ago. Here in the Prescott area, the telltale scars such tectonic battles leave remain preserved in the rocks, allowing you to witness firsthand the construction of a continent.

Mountain of Metal
THE MINES OF JEROME

The historic mining center of Jerome bills itself as the world's largest ghost town. While today this bustling artist colony and tourist destination is anything but deserted, like most mining towns, it has seen its share of boom and bust, and it came perilously close to extinction after its last big mine closed in 1953. At its peak in the 1920s, Jerome's population of 15,000 made it the second biggest settlement in the Arizona Territory. It boasted a school, an opera house, two churches, and numerous illicit saloons. The economy was vibrant and, despite the town's precarious perch on a steep mountainside, a railroad line was pushed through the mountains to bring in goods and cart out ore. The expense involved in such efforts was considered of little consequence, because the town quite literally sat atop a mountain of metal. The town boasted two ore bodies so fabulously rich in copper, gold, and silver that they were classified as "bonanza" grade. It is not uncommon for a modern copper mine to profitably excavate ore that contains a mere 0.5 percent copper. In contrast, Jerome's United Verde Extension mine, also

Jerome in the 1920s, during the boom years —Jerome State Historic Park

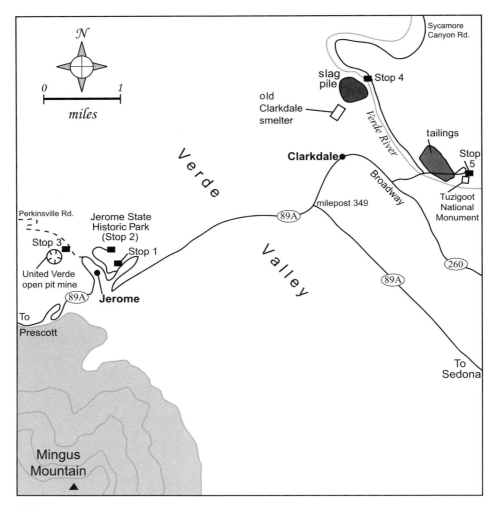

❖ GETTING THERE

Jerome is 56 miles southwest of Flagstaff. Within the town limits, stop 1 is a large dirt pullout and scenic overlook on the north side of Arizona 89A at milepost 345. To reach stop 2, Jerome State Historic Park, a museum, head west (uphill) on Arizona 89A for 100 yards to the clearly marked right turn. Follow this road 1 mile to the paved parking lot. To get to stop 3, return to Arizona 89A and turn right (west). Follow the highway 0.7 mile through downtown Jerome to a hairpin left turn at the fire station and go straight, past the right side of the fire station, on Perkinsville Road. Proceed 0.2 mile to a large dirt area on the left with a good view into the mine. To reach stop 4, return to Arizona 89A, turn left (east) and drive downhill 4.8 miles to a major road junction (milepost 349). Continue straight, following the brown signs to Tuzigoot National Monument. After downtown Clarkdale, the signs to Tuzigoot will direct you to turn right onto Broadway. Follow Broadway for 0.6 mile to a signed left turn. Follow this road across the Verde River and turn left onto Sycamore Canyon Road for 1.3 miles. The slag pile will be visible on the left, across the river, as you descend a hill. Pull over at a convenient pullout. Stop 5 is at Tuzigoot National Monument. Retrace your route on Sycamore Canyon Road for 1.3 miles, then turn left on the road to Tuzigoot. After 0.5 mile, the road passes through orange tailings before ending in the national monument parking lot. The best view of the tailings is from the hilltop Native American ruin inside the monument.

called the UVX or Little Daisy, extracted ores that averaged 10.23 percent copper and, in places, topped 45 percent.

Between 1876, when the first claims were staked, and 1953, when it finally fell silent, the larger of the two Jerome mines, the United Verde, produced almost 33 million tons of ore. The smaller but higher-grade UVX mine, which operated between 1912 and 1938, served up an additional 4 million tons. In today's market, the copper alone from these two mines was worth about 3 billion dollars. Although gold, silver, lead, and zinc were found in much smaller quantities, the deposits were still rich and totaled another billion dollars.

Minerals containing copper and gold are found everywhere in the earth's crust, but usually occur in minute quantities that could never be feasibly mined. Ore bodies are small pockets with such anomalously high concentrations that they can be profitably extracted. The copper deposits beneath Jerome are some of the richest ever found on earth, with up to 720,000 times the standard crustal abundance of these metals. This extraordinary mineral wealth is the legacy of a chain of geologic events that has taken place over nearly 2 billion years of earth's history. Despite their antiquity, these deposits took just 75 years to excavate. The environmental effects of the mining, however, will persist well into the future. Excavation of Jerome's rich ore deposits has shaped the human history of the entire Verde Valley region. The town and its fluctuating fortunes are a humble reminder that much of human history has been influenced by the vagaries of the geologic processes that shape the land we inhabit, form the minerals from which we construct our civilizations, and produce the riches we covet.

Stop 1 is a perfect location for an overview of Jerome's geologic setting. The highly visible white "J" above town is emblazoned upon Cleopatra Hill, the namesake for the Cleopatra formation, the 1,750-million-year-old volcanic tuff that hosted both the United Verde and the UVX ore bodies. Immediately to the right of Cleopatra Hill is the United Verde open pit mine. The United Verde ore body was first tapped through the shafts of an underground mining operation, but in 1905 a large lump of highly flammable pyrite rock (named for the Greek word for fire) at the top of the deposit caught fire. Once started, the fire was impossible to put out in the confines of the mine tunnels and burned for years. The company installed bulkheads to separate the miners from the smoldering rock, but this proved inadequate when water leaked into the mine and flashed to steam, blowing out a bulkhead and killing six miners. By 1917, access was cut off entirely, necessitating excavation of this open pit to reach the tunnels that penetrated 4,500 feet into the mountain's bowels.

The tip of the United Verde ore body was exposed on the northern slopes of Cleopatra Hill, making it easy to spot. In fact, a Spanish colonial

Jerome, looking west from stop 1. Cleopatra Hill has the "J" on its flanks. The United Verde open pit is visible to the right and behind the hill. Miners hit ore in the UVX almost directly below the Powderbox Church, but the head frames that served the mine are to the right, outside the photo. The roadcuts in front of the Powderbox Church consist of Hickey basalt. The Verde fault runs directly through downtown.

expedition noted the presence of the ore when it passed by in 1585, and it is likely that Native Americans had been mining the deposit for centuries before that in search of beautiful green malachite and blue azurite, both copper-bearing minerals.

In contrast, because it is deeply buried and gives no surface hint of its presence, the UVX deposit wasn't found until 1914. To locate it from stop 1, look a good distance to the right of the open pit, due north. A bright white mansion there is the former home of James Stuart Douglas, owner of the UVX mine, and the present-day location of a museum, Jerome State Historic Park (stop 2). Immediately to the left of the mansion rise two head frames, towers that held the cables once used to transport men and equipment into and out of the shafts of the underground UVX mine. The shafts tunneled toward the south to tap the main ore body, which lay more than 1,000 feet belowground, almost directly beneath the small, solitary building surrounded by evergreens known as the Powderbox Church. Immediately to the left of the museum access road, the church was constructed from discarded dynamite boxes, hence its name.

The discovery of the UVX ore is a tale of geologic detective work and sheer luck. It focuses on the Verde fault, the existence of which you can deduce at this spot. Look again toward the massive, reddish orange rocks of Cleopatra Hill, which at 1,750 million years old are among the oldest in Arizona. Then look at the rocks immediately below the Douglas Mansion. The latter are black and bedded—distinctively different from those of Cleopatra Hill. These are basalts of the 15-million-year-old Hickey formation, a relative geologic youngster. How did rocks of such vastly different ages come to lie adjacent to one another? The answer is that a major fault has brought them together.

The Verde fault is the biggest in a family of normal faults that began to slip about 10 million years ago, dropping the land to the east down as much as 6,000 feet below that on the west to form the modern Verde Valley (more about this in vignette 18). The fault runs right through downtown Jerome and is responsible for Cleopatra Hill's steep slopes. The trace of the fault passes between the town's two largest buildings immediately below and left of the whitewashed "J" on the hill. From there it

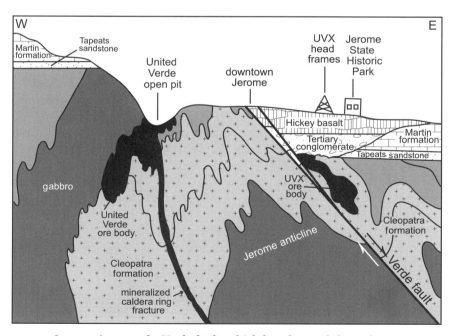

Jerome sits atop the Verde fault, which has dropped the rocks to the east down relative to those on the west. The wavy lines in the rock formations depict how strongly folded these layers are.

runs through the front of the United Verde open pit mine, with the United Verde ore body located on the fault's upthrown (west) side.

Early miners were well aware of the fault and its proximity to the rich United Verde ores. Among others, James Douglas reasoned that the fault must have sheared the ore body in half, suggesting that its eastern side must lie buried beneath the black basalts and underlying white limestones you see below his mansion. In 1912, Douglas purchased the claim where the mansion and the head frames now stand and began to develop the UVX, or Little Daisy, mine. For two years his miners probed deeper and deeper into the earth, excavating nothing but tons of worthless rock. In 1914, on the verge of despair, the miners unearthed a rock comprised of an astounding 45 percent copper. Only in his wildest fantasies could Douglas have envisioned striking so rich a lode. The UVX mine continued to extract amazingly high-grade ore for twenty-four years. Douglas's geologic intuition had given Jerome its second successful mine and made him an even wealthier man than he already was.

The only problem with this story is that Douglas's geologic intuition turned out to be dead wrong. While the miners were quite familiar with the Verde fault, the relationships between the various Precambrian rock units were so hopelessly complex that they were at a loss to fully understand them. It wasn't until the 1980s, long after the ores had been exhausted, that those relationships were fully worked out. At that point it became clear that, while Douglas's hypothesis had been reasonable, the Verde fault had not sliced the United Verde ore in half. Douglas had struck a completely separate ore body that could just as easily never have existed—or been discovered. His bonanza strike was an instance of sheer, dumb luck.

How could two such highly concentrated ore bodies have formed so close to one another? The remarkable story of the ores' formation is vividly illustrated in excellent displays at stop 2, the Jerome State Historic Park museum. The short drive there takes you past the Powderbox Church and the UVX head frames, which rest on the Hickey basalt, evidence that you are still on the Verde fault's down-dropped side. Be sure to leave ample time to explore the museum, whose displays feature fantastic three-dimensional models of the mines and their geologic context, beautiful samples of copper ores, and other exhibits celebrating life in this old mining mecca.

In the historic park's mineral room is a step-by-step chronology of the events that formed Jerome's two ore bodies. Jerome's ore formed a staggering 1,750 million years ago at undersea hot springs nearly identical to vents recently observed along modern mid-ocean ridges. At that time, nothing like our modern continent yet existed. The area that would become northern Arizona was nothing but a tangle of chains of volcanic islands, known as island arcs, and slivers of continents, all reminiscent of the complex geography and tectonics of Southeast Asia today (vignette 16 discusses this complex geometry in more detail). These arcs and microcontinents rode atop a series of small oceanic plates off the shore of the proto–North American continent.

Several of these small plates were subducting beneath their neighbors, creating arc-shaped chains of volcanic islands in the process. The future site of Jerome straddled one such arc, where rhyolite and basalt lavas poured from separate volcanic vents, mingling to build up a volcanic dome. Distinctive, pillowlike structures found in the area's oldest basalts confirm that these rocks were erupted beneath the sea. These pillow basalts form when cold seawater quickly cools the exterior of the lava, forming an oval-shaped crust into which still-molten lava continues to pour (the historic park contains an excellent example). The pillow puffs up until its rind cracks, diverting the lava stream to one side, where it forms an adjacent pillow.

After a brief pause in volcanic activity along this stretch of the arc, a new chamber of gas-rich magma began to work its way up toward the surface directly beneath today's Jerome. This magma's high content of dissolved gases, including water and carbon dioxide, boosted the pressure in the chamber until it exceeded the strength of the overlying rock. The rock ruptured, and the magma burst out in a gigantic caldera eruption (similar to the Peach Springs caldera eruption described in vignette 10), creating the Cleopatra formation, visible in the "J" hill. As the largest and most violent type of volcanic eruptions, caldera eruptions empty the entire magma chamber in a matter of days or weeks, creating a subterranean void that cannot support the weight of the volcano above. The mountain therefore collapses into the void, leaving a giant depression, or caldera, in its stead. This collapse occurs along a circular fault line known as a ring fracture, which outlines the former magma chamber.

In the case of the undersea Jerome caldera, water poured into the numerous cracks and faults created during the explosion. This water percolated deep into the earth's crust, where it mixed with mineral-rich water escaping from another chamber of rising magma. This encounter heated the water to temperatures likely exceeding 660 degrees Fahrenheit, increasing its buoyancy and causing it to rise toward the surface, as well as amplifying its ability to dissolve and chemically alter the rocks through which it passed. In this way, the superheated water scavenged dissolved precious metals such as copper, gold, and silver from the surrounding volcanic rocks, becoming thousands of times more concentrated with them than the host rocks had been.

The caldera's ring fracture provided the perfect conduit for this hot, mineral-laden water to escape from the seafloor. As the water burst from the ring fracture at a hot spring, the surrounding frigid seawater instantly cooled it. Just as cooling sugar water will precipitate out rock candy, the hot spring's load of dissolved minerals immediately solidified, forming black "smoke" that belched from the vent before settling to the seafloor as a blanket of sparkling, metal-rich sediment. Modern submarine hot springs behave in exactly the same way, causing the scientists who first discovered them off the Galapagos Islands in 1977 to christen them "black smokers." Some of the solidifying material coagulates around each vent to form a misshapen metallic chimney that gushes black smoke, giving the vent field the look of some alien industrial complex. The dissolved metals combine with sulfur to form a deposit known as a massive sulfide. Along with the metal-rich linings of the hot spring conduits and the chimneys themselves, a blanket of such sulfides formed Jerome's rich ore bodies.

Due to mixing with shallow groundwater, the water issuing from other oceanic hot springs is sometimes cooler (down to about 250 degrees Fahrenheit), emitting a lighter-colored mineral cloud that earns them

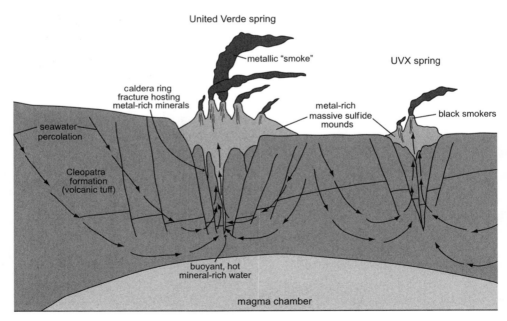

The United Verde and UVX mineral deposits were both formed by the expulsion of superheated, mineral-rich water at submarine hot springs above faults in the Jerome caldera.

Modern black smoker chimney spewing metal-rich "smoke." —Courtesy J. Delaney and V. Robigou, University of Washington

the name "white smokers." The Jerome State Historic Park museum displays both black and white smoker chimneys from the Jerome ore bodies that are nearly identical to those in modern vent fields. The Jerome smokers are among the oldest examples of these features ever found. Such chimneys are of particular significance because many evolutionary biologists believe that life on earth first evolved at submarine hot springs like these. Before the protective atmospheric ozone layer formed around our planet, lethal cosmic rays bombarded shallow water environments. However, deepwater vents would have been safe from this assault, and the chemicals they belched out—for example, hydrogen sulfide—would have provided nourishment for earth's first inhabitants. Archaea bacteria, the most primitive organisms known, thrive at modern submarine hot springs, which lends support to this hypothesis.

Jerome's two ore bodies formed at separate hot springs issuing from the caldera's ring fracture. Eventually, however, the metal-rich water ceased to flow, and the springs became extinct. The next major event in the history of the United Verde and UVX ores occurred 1,700 million years ago, when the volcanic arc hosting the Jerome caldera slammed into the neighboring Prescott arc, followed very soon thereafter by collision with the proto–North American continent itself. These massive collisions so strongly folded the Cleopatra tuff and the hot spring deposits within it that the deposits doubled up on themselves, positioning the twin ore bodies on opposite limbs of a fold known as the Jerome anticline. These types of collisions typically cause large volumes of water to flow through the rocks. It appears that fluids coursing through the ore bodies at this time scavenged copper from nearby lower-grade rocks and, due to a change in water chemistry, redeposited it in Jerome's massive sulfide rocks, further enriching the already fertile mineral deposits.

No rock record exists of the events in the Jerome area over the next 1,200 million years, but by 525 million years ago, the mountain range created by the collision of the Jerome arc with the early continent had eroded down to a level just above the tops of the ore bodies. At that time, northern Arizona was tectonically quiet and hosted a shallow sea. A thin layer of sandstone, the Tapeats sandstone, was deposited on a beach at the edge of that sea, covering the Cleopatra formation. Limestones of the Martin formation and the Redwall limestone later accumulated in successive seas, burying the tuff and the ores more deeply. Still more sediment continued to accumulate here until about 70 million years ago, when a second major mountain-building event known as the Laramide Orogeny swept the region (see vignettes 5 and 9). Compression created new faults, including the Verde fault, which uplifted another high mountain range in the Jerome area. Many of the sedimentary rock layers that had formerly covered the ore bodies eroded off the top of this range, reexposing the Martin and Redwall limestones.

Things quieted down again until about 15 million years ago, when crustal stretching began fracturing western and central Arizona into a series of parallel mountain ranges and valleys, forming the distinctive Basin and Range province (vignettes 1 and 10). In the Jerome area this spasm of tectonic activity first triggered the eruption of the Hickey formation basalts and then, about 10 million years ago, rejuvenated movement along the dormant Verde fault. Moving again, the Verde fault cut the Jerome anticline, exposing the United Verde ore deposit on the flanks of Cleopatra Hill and sliding the UVX deposit 1,000 feet underground. Fluid migration through these new faults further concentrated the copper in the ore bodies, putting the finishing touches on these jackpot deposits.

Instead of slicing the United Verde ore body in half, though, as James Douglas conjectured, the Verde fault simply cut through the Cleopatra

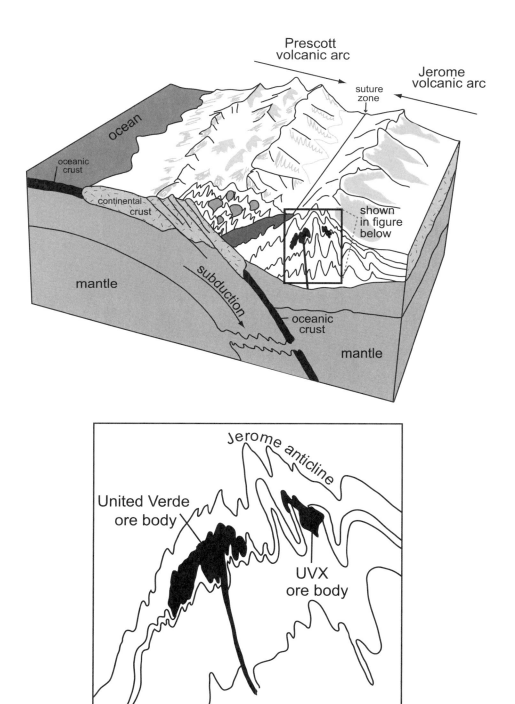

About 1,700 million years ago, the Jerome volcanic arc collided with the nearby Prescott volcanic arc, folding the Cleopatra formation nearly in two to form the Jerome anticline. The United Verde and the UVX ore bodies ended up on opposite sides of this fold.

formation directly along the spine of the old Jerome anticline, fortuitously placing the two distinct hot spring deposits near one another on opposite sides of its accordion-like fold. The museum exhibits at Jerome State Historic Park illustrate each chapter in this tectonic story, including a giant, three-dimensional cutaway model of the Jerome area and some of the UVX discovery ores.

One room in the museum hosts a model of the 88 miles of mine shafts beneath the town. This model was constructed as evidence in the town's 1935 lawsuit against the two companies that owned the mines as a result of a slow-moving landslide that, over the preceding decade, had caused downtown buildings to slip down the hillside. The town blamed the hillside's destabilization on the enormous amount of rock that had been removed from the mountain of metal. Eventually the mines agreed to pay the town $53,500. The sliding ceased with the end of active mining in 1953.

You can get an even more tangible sense of the enormity of mining's impact on this town with an up-close look at the United Verde open pit at stop 3. The lot where you park is built on a huge pile of worthless rock discarded from this gaping hole. Formerly part of the underground mine, the floor of this pit used to lie 300 feet below the surface. Mine shafts penetrate the mountain 4,200 feet beneath your feet, and you can still see the head frames that serviced them. The pit's back wall is formed of hard black gabbro, an igneous rock that is chemically identical to basalt, but which cooled and solidified belowground instead of above. This gabbro was the last igneous rock that formed along Jerome's volcanic arc before it slammed into the neighboring Prescott arc, shutting off all volcanic activity. If you look in front of the wall of gabbro, you can see that portions of the walls on both sides of the pit are crumbly and stained orange. These rocks were sheared and discolored by motion along the Verde fault beginning 10 million years ago. Some of the discoloration also dates from much earlier, caused by mineralization within the caldera ring fracture. The rusty streaks you see are discoloration created by mineral-laden water as it migrated up the old conduit of the ring fracture, in the process enriching these ore bodies 1,700 million years ago.

In the early days, the copper ore extracted from the mine was processed in a smelter right in Jerome. In 1915 the smelter was moved down the mountain to Clarkdale, our next stop, and another was opened nearby at Clemenceau, in the present-day city of Cottonwood. At all of these facilities, the copper was extracted in a multistage process. First, the ore was crushed to a powder and mixed with water to form a sludge. The sludge was apportioned into large flotation vats and vigorously churned, creating frothy bubbles that rose to the surface. Oils added to the vats helped

Two head frames in the United Verde open pit

the copper stick to these bubbles, which were skimmed off the residual sludge, or tailings. The frothy copper concentrate was then smelted, or heated to high temperatures, causing it to melt. Due to their high density, compounds of copper, iron, and sulfur would sink, while lighter materials, known as slag, were removed. On the banks of the Verde River at stop 4 stands the gigantic heap of slag that poured like molten lava out of the Clarkdale smelter.

Further processing of the copper-iron-sulfur compound isolated the copper, resulting in a product that was 99.5 percent pure. In the days of the Jerome and Clarkdale smelters, the sulfur dioxide gas produced by this process was vented out of smokestacks. In the atmosphere, it readily combined with water to form sulfuric acid, the cause of acid rain. Modern mines capture the sulfuric acid to prevent its release into the environment, but in the old days this technology did not exist, and in Jerome, the corrosive fumes killed just about every tree that had not yet been cut down on the formerly wooded hillside.

Stop 5 is the final resting place for the tailings removed from the flotation vats. A wooden pipeline transported this waste in the form of slurry from the Clarkdale smelter to a tailings pond on the floodplain of the

A mountain of slag rises above the Verde River at Clarkdale.

Verde River. These tailings form the barren, orange wasteland adjacent to the hilltop Native American ruin of Tuzigoot National Monument.

Such tailings ponds exist at all copper mines, and they present a special environmental hazard that must be carefully managed to avoid serious ecological damage. The tailings are rich in pyrite and other non-copper-bearing sulfide minerals, which form when elements like iron bond with sulfur in the absence of oxygen. When the sulfide minerals in the tailings pile are exposed to air, the sudden abundance of oxygen causes them to break down and form new compounds, including iron oxide (rust), which gives the tailings their distinctive orange color. The orphaned sulfur becomes free to bond with water, forming sulfuric acid, a strong acid that quickly dissolves other toxic heavy metals like mercury, arsenic, cadmium, and selenium. Tailings piles are thus potential sources of both strong acids and toxic heavy metals that can wreak havoc on local flora and fauna. Another hazard develops if the tailings pile dries out: windstorms can blow toxic tailings dust across a vast area. For these reasons, uncovered tailings must be kept moist, but because that moisture further fosters acid production, the water must be contained within the tailings pond itself and not allowed to escape into adjacent surface water or groundwater.

Tailings from the Clarkdale smelter form an orange wasteland on the doorstep of Tuzigoot National Monument, from which this photo was taken. The largest building, with the prominent white roof (upper right), is the remains of the smelter. Jerome, not in the photo, is to the left.

If you view the tailings from the Tuzigoot ruin, you can easily see the Verde River, one of the few perennially flowing rivers in central Arizona, meandering just 100 feet from the edge of the tailings. No major acid leaks are known to have escaped from these ponds, but the risk is high if the Verde River were ever to experience an especially large flood. Furthermore, the water that has for decades been used to wet the tailings down is sewage water from the city of Clarkdale. For health reasons, the State of Arizona has recently required the city to build a sewage treatment plant and stop discharging their sewage onto the tailings. Because of this, Phelps-Dodge, the current owner of the United Verde mine, is developing a plan to first cap the tailings with impermeable clay to prevent water from percolating through them, then cover that clay with 4 feet of topsoil and reseed the area with native vegetation. Perhaps the tailings will be covered by the time you visit.

The mines of Jerome have shaped the history of the Verde Valley and, indeed, the history of the state of Arizona. Many men spent their entire lives toiling in the mines, and a small handful made their fortunes here. Even today, long after the mines have fallen silent, the residents of the Verde Valley continue to live with the environmental consequences of this extraction and processing of the Jerome ores. The streets of Jerome echo with reminders of the many human lives shaped by two small submarine springs that stopped flowing 1,750 million years ago.

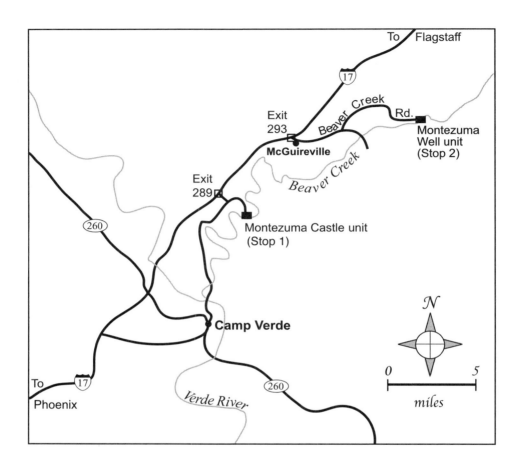

❖ GETTING THERE

Montezuma Castle National Monument is 56 miles south of Flagstaff, just east of I-17 near Camp Verde. Take exit 289 off I-17 and follow the brown signs about 2 miles to the monument visitor center, stop 1. The park's less visited Montezuma Well unit, stop 2, is 4 miles north of Montezuma Castle. Back on I-17, take exit 293 onto Beaver Creek Road, which passes almost immediately through the small town of McGuireville. At a Y-intersection 1.7 miles from the interstate, veer left to stay on Beaver Creek Road. After another 2.5 miles, turn right into the monument just before Beaver Creek Road turns to gravel. The paved road ends 0.4 miles ahead at the parking lot.

A Lake's Legacy
MONTEZUMA CASTLE NATIONAL MONUMENT

The two units of Montezuma Castle National Monument both lie in the Verde Valley, a distinctive northwest-southeast-angling trough framed by precipitous escarpments on either side. The Verde Valley marks the transition from northern Arizona's relatively flat, high-elevation Colorado Plateau to the much lower Basin and Range province characteristic of southern and western Arizona. Moving west, the Verde Valley is the first basin in the Basin and Range. Regardless of which direction you come from on I-17, the abrupt descent will impress you.

The Verde Valley was primarily formed by the large Verde fault, which runs along the basin's southwestern side. As recounted in vignette 17, this fault, active from 10 until 3 million years ago, dropped the Verde Valley down relative to the Colorado Plateau to the east and the Black Hills to the west.

Strange as it may now seem, from the time the Verde Valley began to form until the Verde River was established 2 million years ago, this parched basin intermittently hosted a large lake, ancient Lake Verde. In the center of the lake, layer upon layer of chalky, white limestone was deposited. Montezuma Castle and Montezuma Well, the two signature features of this national monument, both owe their existence to the presence of this limestone.

For most of its life, the bathtub-shaped basin had no outlet for water, so moisture shed off the surrounding highlands pooled on the valley floor to form Lake Verde. During this time, the climate was generally dry, but it veered periodically from wetter to dryer and back again. During wet periods, a perennial lake occupied the valley floor, rimmed by forests hosting such impressive prehistoric animals as mastodons, saber-toothed cats, and camels. Limestone accumulated in the bottom of the lake during these times. However, as the climate repeatedly dried out, Lake Verde slowly evaporated and shrank until it formed a desert playa resembling Death Valley's modern salt flats. At such times, the valley could only host a shallow lake during a few months of especially wet years. During dry spells, mud blanketed the valley floor, and salts accumulated as lake water evaporated under a blistering sun.

With a steady supply of material continually stripped by the elements from the surrounding highlands, the bathtub-shaped basin gradually began to fill and even bury the lower flanks of the adjacent mountains. Eventually, by about 2 million years ago, the sediments had accumulated to a height of over 5,300 feet—the level of the lowest mountain pass on the valley's southern end. At that point, water poured over this breach and flowed south to the Salt River Valley (where Phoenix now lies) and the Gulf of California far below. With this overflow, the Verde River was born, and Lake Verde drained away.

To imagine this process in action, think of filling your bathtub with a few inches of water to form a lake. Repeat the process several times after shoveling some dirt into the tub each time. Eventually, you will fill the tub to the brim with dirt, and when you turn on the tap, the water will spill out of the tub and onto the floor. Your tub no longer supports a lake, but a very efficient river now transfers water from the tub to the lower elevation of your bathroom floor. Erosion long ago removed the original breach, but from Beasley Flat, on the south side of the valley, you can see the modern outlet between Hackberry and Table Mountains, where the valley ends in a narrow gorge.

The birth of the Verde River coincided with the onset of the famous Pleistocene ice age at the beginning of Quaternary time. There were no substantial glaciers in Arizona, but the relatively moist climate the area enjoyed at that time was sufficient to feed a river of significant size. The young Verde River was swift and vigorous where it tumbled into the Salt

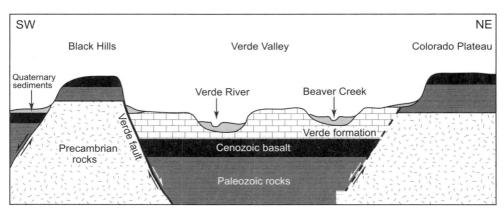

The Verde Valley formed when the Verde fault, on the valley's southwestern side, and possibly a smaller fault on the northeastern side (dashed line), down-dropped this large trough between precipitous 3,000-foot escarpments. Lake sediments of the Verde formation covered the same Cenozoic basalts that cap the surrounding Black Hills and Colorado Plateau. More recently, the Verde River and tributaries cut channels through these lake deposits, exposing them in low mesas.

Alternating layers of hard, white limestone and crumbly, gray mudstones near the I-17 exit to Montezuma Castle

River Valley 1,500 feet below. Both the Verde and its tributaries had the muscle to cut deeply into the soft 3,000-foot pile of lake sediments, dissecting them into a series of low mesas. As you descend into the Verde Valley, look for these low, flat-topped tablelands, the remnants of this once continuous surface. When you pass through large roadcuts, see if you can distinguish the lake's solid white limestone beds from the crumbly gray mudstone and salt layers laid down during drier times. One good place to see these alternating layers is a roadcut on Beaver Creek Road where it intersects Brocket Ranch Road, which you will pass later, 0.8 mile from I-17 on your way to stop 2. But first, proceed to stop 1, Montezuma Castle.

The first European-Americans to see Montezuma Castle believed it to be of Aztec origin and named it for the Aztec king. The castle and Montezuma Well lie in a particularly thick set of limestone layers that were deposited near the center of the lake. In the twelfth century, people of the Sinagua culture recognized a good thing in this durable limestone and an alcove carved by Beaver Creek, a major tributary of the Verde River. The

Sinaguans, contemporaries and neighbors of the more famous Ancestral Pueblo people (formerly called the Anasazi) to the north, built an exquisitely preserved cliff dwelling in this alcove, here at stop 1.

At the visitor center, walk out the back door and stroll along the delightful 300-yard path through the lush riparian corridor of Beaver Creek. Giant, white-barked Arizona sycamore trees line the trail, and the sound of rushing water—a rarity in this state—provides a pleasant background murmur. Montezuma Castle is an arresting sight; high above the river, its ramparts rise like a fortress in a large alcove that dents the tall cliff of lake-deposited limestone. The river, when it lay at that height, carved the alcove from soft rock layers at a bend. Continued downcutting of the river has left the alcove high and dry.

This was an ideal location. Beaver Creek provided reliable water and good agricultural land along its floodplain. The Sinaguans obtained salt from a nearby mine in one of the lake deposit's evaporite layers (they even shored up the mine shaft with timbers, a rare early use of such engineering). The village's limestone alcove provided both shelter from the

Skilled Sinaguan architects built Montezuma Castle in a natural alcove carved by Beaver Creek in limestone deposited in ancient Lake Verde.

elements and protection from enemies. The alcove was naturally cut, readymade for the Sinaguan architects to construct their mud brick walls within—a good thing, given that the limestone is very hard and not easily carved (you can verify its hardness where it is exposed at the end of the trail). The alcove faces east, so it benefited from warm morning sun. By the early fifteenth century, however, the Sinaguan people had abandoned it and other settlements in the Verde Valley for reasons archaeologists don't fully understand. The Verde Valley's ancient lakebed passed into the hands of the Yavapai Apache people, who, with later settlers from Europe, still occupy the area today.

Less visited than the castle, the Montezuma Well unit of the Montezuma Castle National Monument, stop 2, has a peaceful, contemplative ambiance. What the Well lacks in archaeological drama, though, it more than makes up for with its singular geologic feature, a classic sinkhole.

To reach the Well from the parking lot, walk up the long flight of stairs next to the ranger cottage. From sparse desert vegetation you will emerge into an altogether different, greener world. Ringed by lush grasses and

An oasis for wildlife and a stopover for migratory birds, Montezuma Well formed from the collapse of a limestone cave's roof.

reeds, a lake lies in the bottom of the circular sinkhole. A more anomalous sight is hard to imagine. A few low Sinaguan dwellings and granary walls occupy recesses in the sinkhole's steep walls.

Because sinkholes form most readily in limestone, the legacy of ancient Lake Verde is again apparent. Aerial photographs reveal that the Well lies in the center of an especially thick, mound-shaped chunk of limestone. A series of springs issued from the floor of Lake Verde, precipitating limestone as they hit the lake water. Just such a process also occurred in California's Mono Lake, creating the weird and wonderful tufa towers for which that lake is famous. Recent research at the University of New Mexico shows that Lake Verde springs were fed by a mixture of shallow, rain-derived groundwater and a trickle of much deeper water from the earth's mantle. The mantle-derived water brought with it an abundant supply of carbon dioxide, which aided greatly in forming Montezuma Well's thick limestone mound.

Sinkholes, which form when the roof of a shallowly buried cave collapses, are common features of limestone terrain in humid areas. But how do caves form, and why do some eventually collapse to form sinkholes? The key is the interaction between water, which is present everywhere beneath the ground, and limestone bedrock. Some distance below the earth's surface, soil or rock material is completely saturated with water, like a sopping-wet sponge. The flat top of this saturated area is called the groundwater table. In areas with limestone, caves form below the groundwater table where water dissolves the limestone bedrock. That is because limestone is particularly susceptible to dissolution by acids (geologists test for limestone using hydrochloric acid, the origin of the popular term "acid test"). Rainwater is slightly acidic due to its natural combination with carbon dioxide in the atmosphere, forming weak carbonic acid. Groundwater picks up additional acids from plant roots and soil bacteria, making it even more corrosive to limestone. The rock dissolves along joints, or cracks, and bedding planes—weaknesses the water easily exploits. Caves slowly develop along these weaknesses into a complex series of large rooms connected by narrow passages. As long as the chambers are filled with water, water pressure helps support the cave roof. However, if the climate dries out (or if humans pump the groundwater vigorously), then the water table drops, leaving a cave filled with air. If the roof of the cave is thin, it is now unsupported and liable to collapse, forming a sinkhole. Montezuma Well formed when the roof of a circular cave collapsed.

As you gaze at this green oasis, its stark contrast with the surrounding desert vegetation raises an interesting question. How, in this dry basin, does the water stay in the bottom of the hole? In general, lakes form on

HOW SINKHOLES FORM

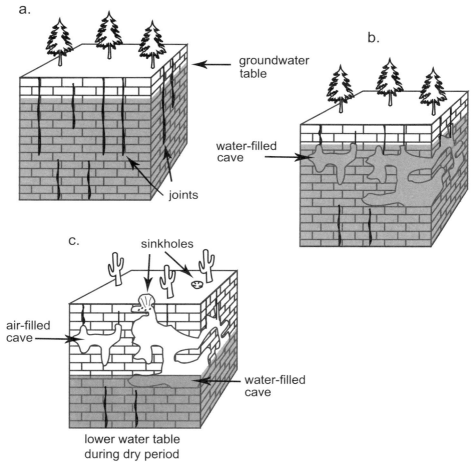

the floors of sinkholes when the groundwater table lies *above* the bottom of the hole. At this location, however, the main water table lies *beneath* the floor of Montezuma Well.

To look for an answer to this puzzling question, take the paved path from the overlook. After 10 yards, follow a branch to the left, to Swallet Ruin, as it descends a series of stone steps into the Well itself. You will pass more Sinaguan ruins just before the 150-yard path ends at a natural underground outlet, called a swallet, that drains the Well. The amount of water entering the lake is balanced by water disappearing into this miniature cave.

To see where this water goes, retrace your steps to the rim of the Well and continue walking counterclockwise around the sinkhole perimeter. About 200 yards past the Swallet Ruin trail junction, the path leaves the rim and passes the foundations of several Sinaguan structures before descending 130 yards to another fork in the trail. Follow the left branch down a series of rock steps to Beaver Creek. As you descend the last several steps, where the trail comes closest to the limestone cliff on the left, keep a sharp eye out for small leaf impressions in the rock, testimony to the forest that grew along the shores of Lake Verde a few million years ago. Please don't touch these fossils, as they are easily damaged.

The path dead-ends at a magnificent sycamore tree and a beautiful spring where the Well water reemerges seven minutes after dropping into the swallet you saw earlier. The water flows out of the spring and into an ancient Sinaguan irrigation canal before flowing into Beaver Creek. At this spot you are standing below the bottom of the lake that occupies the

The ancient Sinaguan people diverted the waters of the spring that drains Montezuma Well into a canal for irrigating their crops.

Well, but clearly the water table still lies below your feet. What, then, is keeping the water in this sinkhole crater?

Although water readily flows through many soils and different types of rocks, there are a few it cannot pass through easily. Mudstones, for example, are composed of clays, platy minerals that overlap one another like shingles on a roof. Water passes through this shingling extremely slowly, causing water to pool above it. Mudstones deposited during a dry period in ancient Lake Verde's history formed just such an impermeable layer here. Except at the one swallet you saw, this underlying mudstone bed prevents water from draining away, as it has at Arizona's other sinkholes. The pooling of water above this mudstone creates a small, localized patch of water that hydrologists call a perched water table, because it lies above the main water table here.

While you linger at the spring, take a look at the numerous holes in the limestone cliff above you. Many are lined and partially clogged with large, white crystals of the mineral calcite. The holes are either limestone casts that once encrusted plant roots (now decayed), or they formed as percolating acidic groundwater dissolved the limestone. In either case, each hole became a conduit for groundwater escaping from the rock face. The groundwater carries with it dissolved calcium and carbonate from the limestone it has consumed. When chemical conditions are

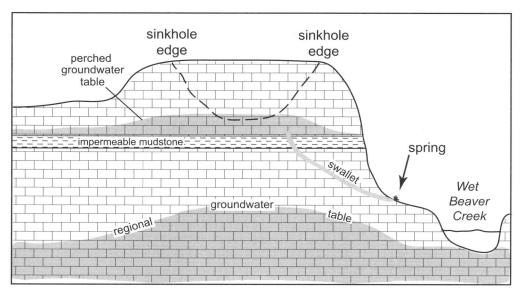

Formation of the perched water table at Montezuma Well

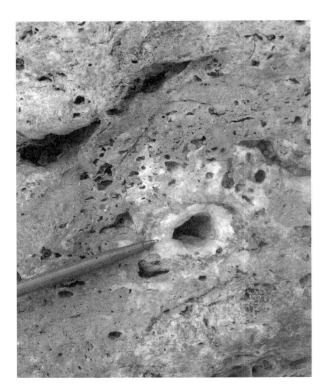

Calcite deposits in the limestone wall above the spring; pen for scale

right, the dissolved materials come out of solution and reform as crystals of calcite, the mineral that makes up limestone. This precipitation of calcite is what forms long, tapering stalactites and other beautiful cave decorations. Here, the cavities through which the water seeped gradually narrowed as they filled with white calcite, like arteries clogging with plaque.

For those who enjoy a good puzzle, there is one final bit of detective work to be done. Low on the cliff just above the spring you can find a large, partially rounded block of black basalt, a volcanic rock, embedded in the light gray sedimentary limestone wall. If you examine the area more closely, you will spot dozens of smaller black pebbles scattered throughout the nearby limestone. The black basalt is derived from lavas deposited along the edges of the Verde Valley. How, then, did this basalt boulder and its smaller satellites get here? Although we aren't aware of any formal geologic study that has answered this question, we have a few ideas based on local evidence.

As indicated by its substantial thickness, the limestone here was deposited near the center of Lake Verde when it was perennial and fairly deep. That would rule out transport of the basalt boulder by a river, which would deposit its load at the lake's edges, not in its center. In addition, a river would bring with it lots of boulders and cobbles that would form a discrete layer, not just the few scattered bits and pieces present here. The lavas from which this basalt is derived were erupted during the lake's life span, but if a lava flow had entered the lake here, it would likely form a continuous basalt layer. No glaciers were active in these parts that could have moved the basalt boulder here, and even if they had been present, they too would have left a thick layer of debris instead of these isolated bits.

What other mechanism could account for the presence of the black rocks? Our best guess is that the basalts were transported to the middle of the lake in the root-ball of a large tree that fell into the lake or was swept into it by a flooding river. As the tree floated on the lake, the soil on top of the root-ball dried and sloughed off. Chunks of the basalt rock trapped in the soil plunged to the lake bottom below, to be entombed in accumulating limestone. A modern example of similar "debris rafting" occurs when icebergs calve off of tidewater glaciers. They drift through the oceans and slowly melt, dropping scattered aggregations of encased rocks onto the deep, muddy seafloor.

When you leave lush Montezuma Well, you quickly return to the parched desert that typifies most of the Verde Valley. It is not hard to understand why the Sinaguan people located villages at Montezuma Well and Castle—oases created by the erosion of limestone left by an ancient lake.

A boulder of basalt embedded in the limestone above the spring. How did it get there?

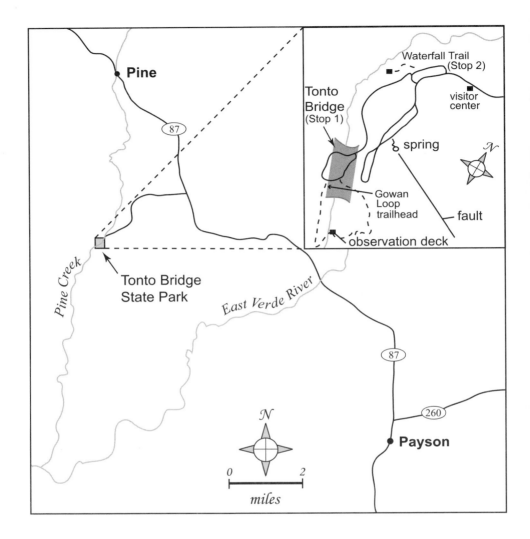

❖ GETTING THERE

Tonto Bridge State Park is southeast of Flagstaff between the towns of Pine and Payson. From Arizona 87, 0.2 mile north of milepost 263, turn west onto the clearly marked road to the park, which descends steeply off the mesa into the canyon of Pine Creek. Park in the bridge parking area to access stop 1, the Gowan Loop Trail, which is wheelchair accessible to viewpoint #4 and moderately steep and well maintained beyond that. In 2006, rockfall closed the eastern portion of the trail beyond viewpoint #4. Therefore, although it may be possible to hike the entire loop by the time you visit, the description here guides you out and back along the west leg of the loop. Allow an hour for the half-mile hike. To reach stop 2, the Waterfall, walk or drive to the Waterfall Trail parking lot. The moderately steep, 100-yard-long Waterfall Trail leads down a series of steps.

19
A Bridge to the Past
TONTO BRIDGE STATE PARK

Massive, soaring Tonto Bridge, the largest natural travertine bridge in the world, lies at the bottom of rugged Pine Creek Canyon in Tonto Bridge State Park. In most ways, the canyon resembles other steep, narrow defiles slicing through the wild country below the Mogollon Rim—the band of cliffs marking the southern edge of the Colorado Plateau. But wide, flat, pastoral meadows in this short stretch of the canyon distinguish Pine Creek from its neighbors. These acres of vibrant green fields crisscrossed by babbling brooks actually comprise the top of Tonto Bridge. For hundreds of thousands of years, a mass of travertine, a type of limestone, slowly crystallized out of spring waters here, plugging and leveling this part of the draw. This was step one in the formation of the bridge.

Pine Creek flows beneath Tonto Bridge, the largest natural travertine bridge in the world.

Travertine is limestone that formed on land rather than in the more common setting of an ocean. It forms the stalactites, stalagmites, and other features that make caves so beautiful and has been a favored building stone for millennia. Many of the famous ruins of ancient Rome are built of travertine, and it is still commonly used today for decorative floor tiles.

From its junction with Arizona 87 to the lip of the Pine Creek gorge, the park access road traverses the flat top of Buckhead Mesa, which is capped by relatively young (13- to 10-million-year-old) black basalt flows. Pause for a moment at the rim of the gorge and notice that the canyon's opposite (west) wall consists of purple, westward-tilting rock layers. These are bands of quartzite, a very hard stone created by the fusion of quartz grains when sandstone metamorphoses. These quartzite layers belong to a very old group of rocks known as the Mazatzal quartzite (despite the spelling, Arizonans pronounce this MA-ta-zal). This rock has been dated at 1,700 million years old. Scanning the gorge's near (eastern) wall, you will notice that these purple, cliff-forming layers are completely absent.

As you drive slowly down the steep road to the park, look at the roadcuts along the way. The highest cuts feature the same 12-million-year-old black basalt that caps Buckhead Mesa, underlain by slightly older river gravels

Precipitation of travertine formed the flat floor of the otherwise precipitous Pine Creek gorge. The gorge's left (east) wall is composed mostly of horizontal Paleozoic limestones, whereas the right (west) wall consists of bullet-hard, steeply tilted layers of Precambrian Mazatzal quartzite.

loaded with rounded cobbles. Below the gravels, the rock turns to creamy gray limestone belonging to three different formations 300 to 380 million years old. Unlike the tilted quartzite across the canyon, rock layers on this side lie nearly horizontal. Such dramatic change in the composition and tilt of rock layers over so short a distance often signals a fault, but extensive field mapping reveals no fault here. Instead, Pine Creek has simply exploited the contrast in strength along the ragged contact between the quartzite and the limestone to carve this striking gorge.

To get your first glimpse of Tonto Bridge, walk the very short paved portion of Gowan Loop Trail to viewpoints #3 and #4. Viewpoint #4, in particular, affords spectacular views of the bridge from a perch on its top.

The view of Tonto Bridge from the top is impressive, but the view from below provides a whole new perspective on its architecture. Getting to that vantage point on Gowan Loop Trail affords views of the rocks on which the bridge is built, allowing you to reconstruct the sequence of geologic events that led to the creation of this natural wonder. Backtrack to viewpoint #3 and take the left (west) fork of the trail to the restrooms west of the bridge parking area. The Gowan Loop Trail turns to dirt here. Follow the descending dirt trail south for about a quarter mile to the observation deck below Tonto Bridge.

From the deck, you can appreciate the immense size of the bridge. It rises 183 feet above your head, and the arch span is 150 feet wide. If Pine Creek isn't flowing too high, you can walk the entire 400-foot length of the tunnel the creek has carved through the travertine to create the bridge.

Here at the observation deck, the very young, white travertine rock of the bridge rests on much older, hard, pink volcanic rock flecked with small, white, rectangular feldspar crystals. Recent dating of this rock, known as the Haigler rhyolite, reveals that the volcano that erupted it was active an astounding 1,709 million years ago.

Once you have soaked up the view of Tonto Bridge, backtrack across the wooden bridge over Pine Creek to begin the ascent out of the canyon. Almost immediately, the rocks change from red rhyolite to purplish Mazatzal quartzite, the same rock you glimpsed at a distance on the drive into the park. The lowermost layers of this formation contain distinctive round cobbles and small boulders; their large size testifies to the swiftly flowing rivers that deposited the sediment of which the Mazatzal quartzite is formed. Through dating of one thin bed of rhyolite (not visible from the trail) sandwiched between these cobble-bearing layers and sandier layers above, scientists have determined that these rocks were deposited approximately 1,700 million years ago.

View from the observation deck at the base of Tonto Bridge. The ancient red Haigler rhyolite upon which Tonto Bridge is built appears dark and compact, while huge travertine blocks eroded from the bridge are light-colored and pockmarked.

Now that you've acquainted yourself with the entire cast of rocks in our story, you're ready to piece together the geologic events that led to the formation of Tonto Bridge. The story begins 1,709 million years ago, when a chain of volcanic islands that would later form the core of northern and central Arizona was erupting across 1,000 miles of ocean (vignette 16). One volcano in this chain erupted the Haigler rhyolite.

Sediment eroded from the volcanic mountains was deposited on an alluvial fan that covered the rhyolite. Onto this erupted one last, small rhyolite flow. Later events cemented these sediments into rock, squeezing and metamorphosing that rock into the Mazatzal quartzite. Those same forces tilted the quartzite and the underlying rhyolite down to the west, as you see them today. We'll tell this tale in more detail in vignette 20.

A great deal more happened to these rocks over the next 1,300 million years, including hundreds of millions of years of relentless erosion that flattened the area and erased much of the rock record. Amazingly, though, a number of small quartzite ridges resisted this erosional onslaught. When a shallow sea flooded the flat expanse about 380 million years ago (vignette 20), they stood as islands above the waves. The quartzite ridge on the west side of Pine Creek canyon was one of these. From 380 to 300 million years ago, the lime that settled out of the seawater accumulated along its flanks.

Much later still, geologically young gravels and basalt were laid down on top of the thick stack of limestone (in the east) and quartzite (to the west). Sometime after 10 million years ago, Pine Creek began carving a gorge through the basalt, and it quickly encountered the ultrahard quartzite. The same toughness that enabled the quartzite to resist the earlier extended period of intense erosion allowed it to rebuff all attempts by Pine Creek to carve through it. Following the path of least resistance, the creek gave up. Instead, it slid down the flank of the old, buried island, exploiting the relative weakness of the limestone and carving the modern gorge. Within the last few hundred thousand years, the creek cut clean through the stack of limestone and began working on the rhyolite you saw along its banks.

Meanwhile, water seeped into the ground from rainstorms up on Buckhead Mesa. It found its way through the limestone along faults created during a spasm of regional crustal stretching that accompanied the eruption of the mesa-top basalt. This water accumulated underground, saturating the rock until it resembled a sopping-wet sponge. The flat top of this saturated area is called the groundwater table, and when the ever-growing gorge of Pine Creek sliced down deep enough to intersect it, water began to leak from the ground in a spring at the tip of one of the faults. Here it encountered a muddy layer in the limestone that was less permeable and blocked the water from percolating deeper through the rocks. This springwater created the massive plug of travertine in which Tonto Bridge was carved.

Why would the springwater leave behind a mass of travertine? Let's continue this discussion at stop 2, the Waterfall, at the end of the 100-yard Waterfall Trail. In this idyllic hanging garden on the banks of Pine Creek, fresh springwater cascades down a travertine cliff verdant with maidenhair ferns and other water-loving vegetation. The trail ends at an alcove covered with little travertine marbles that precipitate from seeping water. Note the thin crust of travertine on the living ferns.

When groundwater travels down through thick sequences of limestone, as it does here, it dissolves part of the bedrock. More than most other types of rock, limestone is susceptible to dissolution. Rainwater is usually slightly acidic because carbon dioxide in the atmosphere combines with water vapor, forming small quantities of carbonic acid. As rainwater percolates through the ground, it typically becomes even more acidic as it encounters organic materials secreted by plants. The

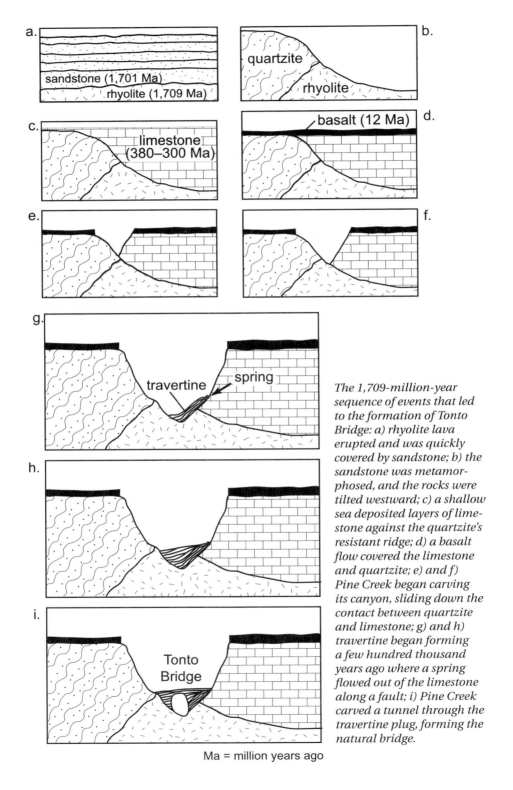

more acidic it becomes, the more limestone it dissolves. This process ultimately forms both caves and sinkholes (vignette 18).

The area's thick sequence of limestone provides plenty of raw material for the groundwater to dissolve. As the water does so, it picks up ever higher concentrations of dissolved calcium and carbonate ions from the breakdown of the limestone. As long as this mineral-laden water remains underground, it remains under high pressure, and the ions stay dissolved. But when the groundwater trickles into the open air at a spring, it loses pressure, and the carbon dioxide bubbles away just as it would if you cracked open a can of soda. After the loss of carbon dioxide, the calcium and carbonate ions recombine into solid calcium carbonate, or travertine.

Recent research at the University of New Mexico has revealed that the water emerging from the spring at Tonto Bridge actually derives from two separate sources. Most of it is rainwater that percolates through the limestone rock, as described above. But a small portion has worked its way to the earth's surface from deep in the earth's mantle. This mantle-derived water, especially rich in carbon dioxide, greatly enhances the springwater's ability to produce travertine.

Some travertine precipitates out at the spring, but crystallization continues downstream as the babbling water continues losing carbon dioxide. Agitated water at waterfalls and rapids accelerates the process. Organic matter also contributes to travertine crystallization. Bacteria

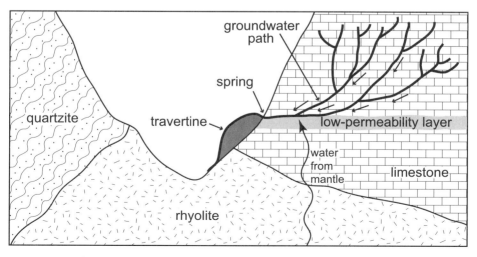

Travertine formation at Tonto Bridge. When the groundwater emerges at the spring, it loses carbon dioxide, causing a chemical reaction that crystallizes travertine.

Travertine forming at the Waterfall, aided by bacteria living on the ferns

that colonize ferns, leaves, and twigs enhance precipitation through their natural metabolic processes. This explains the travertine on the plants here.

Over the millennia, the growing mound of travertine pushed Pine Creek farther west, first pinning it against the hard Mazatzal quartzite, then plugging it by stretching from wall to wall. With nowhere else to go, the water began to seep beneath the plug along the weakness formed by the contact between the rhyolite and the travertine. Unlike the waters coming from the spring, those of Pine Creek were not loaded with calcium and carbonate ions, so they could dissolve the travertine, hollowing a tube out of the base of the plug. This tube quickly became the new path of Pine Creek, and Tonto Bridge was born. To this day, the waters of Pine Creek continue to expand the size of the hole while the springwater adds travertine to the top of the bridge.

Travertine mounds are common in regions with thick limestone sequences. Northern Arizona is one such region; nearby, you can also find spring-formed travertine mounds along the East Verde River and

Fossil Creek and in the Grand Canyon. But nowhere else in the world were geologic circumstances exactly right for producing a travertine bridge as large or dramatic as Tonto Bridge. If not for the thick pile of marine limestone, the faulting that channeled the water to the spring, the presence of a deep gorge that intersected the groundwater, the hard quartzite buttress that trapped the creek and forced it beneath the plug, and the rhyolite contact along which it seeped, there is a good chance this bridge would never have formed. Here in the obscure shadows of a small, little-known central Arizona gorge, Tonto Bridge is a world-class geologic marvel.

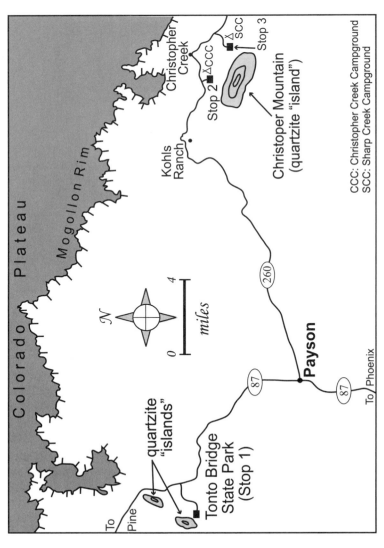

◈ GETTING THERE

Stop 1 is the Gowan Loop Trail in Tonto Bridge State Park. See "Getting There" on page 260 for directions. To reach stop 2 from the park, return to Arizona 87 and turn right. Travel 14 miles to Payson and turn left (east) onto Arizona 260. Drive 20 miles and turn right 0.3 mile past milepost 272 down the short access road to Christopher Creek Campground. The campground is closed in winter, but you can still park and walk a few hundred yards to the areas of interest discussed in the text. Stop 3 is at Sharp Creek Campground (also closed in winter), which is 3.3 miles farther east on Arizona 260; turn right 0.6 mile past milepost 275 onto the access road. Follow signs to the group sites and park in the lot for group site 3.

CCC: Christopher Creek Campground
SCC: Sharp Creek Campground

20

Islands in Time
THE MAZATZAL QUARTZITE

Geology textbooks teach that sedimentary rocks are laid down horizontally (the law of original horizontality) and uniformly over vast, flat expanses of terrain (the law of lateral continuity). These are two of the three "laws" of sedimentary rocks (the third is the superposition of younger layers atop older ones). Postulated and published by Nicolas Steno back in 1668, these principles formed the foundations upon which the science of geology was constructed.

Of course, Steno never believed that sediment layers accumulated perfectly horizontally, and obviously even the largest sediment deposits had to end somewhere. In formulating his laws, he meant only that, at the landscape level, sedimentary rocks were laid down as extensive sheets of material much thinner than they were wide and at a very small angle to the horizontal. On a localized scale, however, Steno's laws are not rigorously true. Occasionally, the presence of a nonhorizontal landscape feature such as a riverbed, a hillock, an offshore island, or a depression may actually be preserved, entombed through the ages by a soft rain of sediment. In such cases, careful study of the surface of contact between the rocks can enable geologists to reconstruct these ancient features tens of thousands of millennia later.

Central Arizona's Mazatzal quartzite (pronounced MA-ta-zal) offers just such a special case. A metamorphosed variety of sandstone, quartzite is exceptionally resistant to erosion because it consists almost exclusively of the mineral quartz. Due to its extreme hardness, quartz is almost impervious to both physical and chemical weathering processes. When the quartz grains fuse together into quartzite's seamless matrix, the resulting rock is nearly indestructible.

Because of this toughness, the Mazatzal has survived through the ages despite multiple exposures on the earth's surface. If you had the chance to sit down over a cup of coffee with a rock unit and listen to its tales of days gone by, the Mazatzal quartzite would be the unit to invite; it has witnessed just about every major landscape change in Arizona's long history. Extensively exposed around the town of Payson, this rock of ages has formed the substrate for generations of younger rocks. Examining its contacts allows us to sketch northern Arizona's changing geography and surface conditions in detail.

At stop 1, walk down the Gowan Loop Trail to Pine Creek at the base of Tonto Bridge (the formation of which is described in vignette 19). The east side of the man-made bridge crossing this creek is anchored in a pink volcanic rock known as the Haigler rhyolite. On the creek's west side lies a stack of reddish purple, layered rocks: the base of the Mazatzal quartzite.

You can closely examine sedimentary features preserved in the lowermost layers of the Mazatzal rocks in the first outcrops you encounter above the wooden bridge. These layers consist of conglomerates brimming with large particles embedded in a finer-grained matrix. Some of these clasts, as geologists call them, are the size of small boulders. Clearly, it would take a lot of energy to move pieces of rock this large.

Notice that the vast majority of clasts consist of red rhyolite identical to the rock exposed on the east side of the bridge. From these clues, you can tell the Mazatzal sediments were moved here by some high-energy process that was chiseling out rhyolite boulders from adjacent, tectonically uplifted areas. The preponderance of rhyolite in the clasts indicates that those highlands must have been very close by; clasts from more distant locations would consist of a greater diversity of rock.

Conglomerate in the lowermost layers of Mazatzal quartzite on the west side of Pine Creek along the Gowan Loop Trail; handheld dictaphone for scale

The Mazatzal quartzite was deposited on a series of alluvial fans shed from a tall mountain range, similar to these two smooth, triangular fans at the base of the Panamint Mountains in Death Valley.

These and other characteristics of the conglomerates have led geologists to conclude that the rocks you are looking at were laid down near the apex of an alluvial fan bounded to the north by a tall, steep mountain range. Numerous similar fans exist throughout Arizona and other areas today. They form right at the mountain front, where steep canyons reach the flatter valley floor. In the mountains, streams flow swiftly enough to carry particles as large as boulders, but the abrupt change in gradient when streams reach the valley floor causes them to precipitously lose velocity, and with it, their ability to transport such large blocks. They unceremoniously dump boulders and cobbles, building a fan-shaped pile of very coarse-grained sediment sloping away from the mountain front at angles typically between 1 and 4 degrees.

As you continue up the trail on the west side of the creek, watch for bedding planes separating one sediment layer from the next. These planes tilt steeply down to the northwest at angles exceeding 30 degrees. Steno's law of original horizontality tells us that the beds' present angle was not depositional, but rather caused by tectonic events after the original sediments were buried deeply enough to be turned to stone.

The Mazatzal layers here are tilted west, so you encounter increasingly younger rocks as you hike west up a series of switchbacks. As you climb, the cobbles and boulders of the lowermost layers give way to pebbles and intervals of pure sand festooned with small crossbeds—divergently angled laminations that form when currents sweep sediment around. This decreasing particle size actually records the wearing down of the mountains through time. With the mountain front eroding northward, the alluvial fan's apex shifted away, and this area became its middle reaches; floods of waning velocity delivered plenty of sand and pebbles, but nothing larger. The canyon wall northwest of the trail displays additional layers of tilted Mazatzal rocks. Here, even finer-grained sediments—metamorphosed sandstone in the lower layers, mudstone in higher ones—record the further erosion of the mountains until they were gone completely.

We now know that an impressive range of mountains once stood here. But how long ago? Another feature of the Mazatzal quartzite has allowed geologists to precisely date the deposition of this ancient alluvial fan. It's a single pink rhyolite bed, identical to the Haigler rhyolite below, sandwiched between sandstone layers above the trail (but not visible from it). Geologists have dated this rhyolite bed and found that it was erupted 1,701 million years ago. The presence of this rock near the base of the Mazatzal indicates that the conglomerate began to accumulate just slightly earlier than that, probably around 1,703 million years ago. This makes it one of the oldest rocks in Arizona.

At that time, southern Arizona didn't even exist, and the land that is now northern and central Arizona was a complex maze of volcanic island chains, reminiscent of modern Southeast Asia, stretching across a thousand miles of ocean (see figures on pages 222 and 223). These volcanic arcs, as such features are called, formed in response to subducting slabs of oceanic crust. Attached to one slab of downgoing crust was the early North American continent. For the preceding 150 million years, subduction had dragged the continent ever closer to the volcanic island chains, including the Jerome and Prescott volcanic arcs. These two arcs, along with many similar ones in the region, were known collectively as the Yavapai arc. A collision with the continent was inevitable. (Vignette 16 tells this story in full.)

Within the volcanic arc, generations of volcanoes evolved and died. One of these, a large, explosive caldera, erupted the Haigler rhyolite you saw by the creek. At about the same time, 1,709 million years ago, the early North American continent finally smashed into the Yavapai arc. The impact shut off the volcanism and thrust up a major mountain range. In the meantime, triggered by local extensional forces, a small basin had formed in the volcanic arc's interior. Into it poured the earliest sediments shed off of the new coastal range; these would become the Mazatzal quartzite. The range had quickly begun to erode; the mountains would be stumps of their former selves within about 50 million years. The lone rhyolite bed above the trail seems to mark the arc's last gasp of volcanic activity.

In the basin, the sediments that would form the Mazatzal quartzite were progressively buried under later sediments, a pile easily several miles thick. This deep burial compressed the sediments into sandstone, then went a step farther, metamorphosing it into the tough-as-nails Mazatzal quartzite. Luckily for us, though, the metamorphism was not intense enough to obliterate the rock's original sedimentary features, preserving clues with which we have reconstructed its birth.

After its burial and metamorphism deep beneath the surface, the Matatzal quartzite lay in wait for a tectonic upheaval to send it upward. It didn't have long to wait. By about 1,650 million years ago, another collision in southeastern Arizona thrust, twisted, tilted, and buckled the Mazatzal quartzite, incorporating it into Arizona's second great mountain range. This upheaval has been named the Mazatzal Orogeny after the Mazatzal Mountains Wilderness west of Payson, where evidence of it is well exposed in outcrops. It was this event that caused the tilt to the layers you have observed at Tonto Bridge.

After this second uplift event ended about 1,600 million years ago, things quieted down in Arizona for a good long time—about 400 million years. During most of this interval, the Mazatzal was probably buried under another blanket of sediments. These rocks are now gone, stripped away by erosion prior to 1,200 million years ago (we'll examine evidence for this at stop 3.) After erosion removed the overlying rocks, it began whittling away at the Mazatzal itself. The quartzite, however, proved too tough to wear away completely. Chunks of it stood high, forming islands in a sea that began encroaching from the south. This is a story that the Mazatzal would repeat time and again. Although we can't see the details of what transpired 1,200 million years ago at Tonto Bridge, we can see evidence of two more recent episodes in which the Mazatzal formed hills in an otherwise subdued landscape.

From the end of the loop trail, walk over to the grassy platform and parking lot atop the natural bridge for a good view of both of the Pine Creek gorge's impressive walls. The west wall displays a ragged outline formed by steeply tilted layers of Mazatzal quartzite. Not much rock pokes out from the canyon's thickly vegetated east wall, but outcrops you can spot are horizontally layered and look quite different from the quartzite you examined by the bridge. As we saw in vignette 19, the canyon's east and west walls bear little geologic resemblance to one another. As you drive up the steep road out of the canyon on your way to stop 2, you'll get closer looks at the canyon's east wall. The roadcuts display mostly horizontally bedded limestone belonging to several formations of Paleozoic age, 380 to 300 million years old. Like most limestones, they were deposited in a sea that intermittently covered this area during that interval.

Using Steno's law of lateral continuity, mentally extend the east wall's horizontal layers across the gap formed by the canyon to where they would butt up against the tilted Mazatzal quartzite. There, all lateral continuity ends. There can be only two explanations for this abrupt juxtaposition of rocks of radically different ages. The first is that a fault running down the canyon shifted these layers into contact with one another. However, thorough geological mapping has shown that no such fault exists here. So the Mazatzal quartzite must have protruded from the seafloor as an island while the younger limestone layers accumulated around it, lapping onto its shores. Amazingly, the ridge of quartzite that forms today's canyon wall was, 380 million years ago, a small, mountainous island roughly 3 miles long and 2 miles wide that boldly protruded 700 feet above the waves of an encroaching sea.

After the Mazatzal's first exposure on the earth's surface 1,200 million years ago, it was buried by sediment until tectonic movements about 750 million years ago triggered a new cycle of erosion that unearthed it again. By 525 million years ago, a sea had crept across the flattened, eroded landscape from the west. By 380 million years ago, the sea level

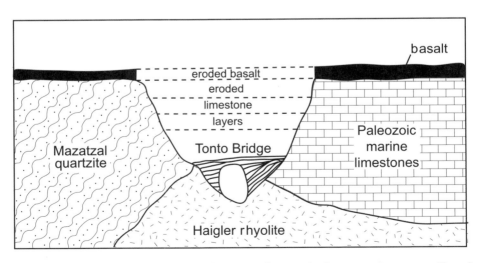

Pine Creek's canyon consists of tilted Mazatzal quartzite layers on its west wall and horizontally bedded Paleozoic limestones on its east wall, all capped by much younger basalt. The limestone layers used to extend across the canyon, abutting the quartzite.

had risen high enough that lime began to accumulate on the flanks of the Matatzal island, where the east wall of Pine Creek canyon exists today. Sediments accrued for another 80 million years, rising ever higher up the island's flank, until they finally buried the island altogether about 300 million years ago.

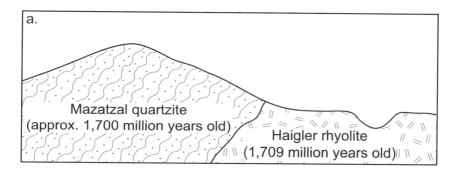

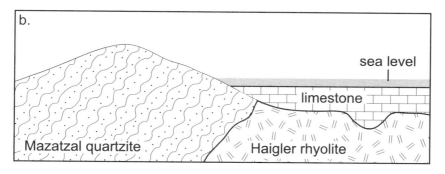

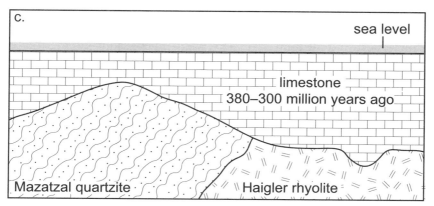

The resistant Mazatzal quartzite formed hills (a) that stood above the rising sea (b). Limestone deposited in that sea eventually buried the island (c).

Driving up out of the canyon, a mile past the visitor center the limestone roadcuts give way to loosely cemented conglomerates. These are gravels deposited by a river that flowed here between 25 and 13 million years ago. A further 0.2 mile up the road, gravels give way to 12-million-year-old black basalt, which forms the canyon rim and the adjacent tableland. If you pull over at the brake check lot on top of the mesa and walk the short distance to the rim, you get an excellent view of the canyon and the very different rock units that make up its walls. The basalt you are standing on here is the only unit the two canyon walls have in common; you can see it forming the crest of the western wall just as it covers the flat mesa top here. The identification and dating of these last two rock layers, the river gravels and the basalt, help us fill in the next few chapters in the Mazatzal quartzite's story.

After lime finally buried the quartzite island now visible in the western wall of Pine Creek's canyon 300 million years ago, sediment continued to accumulate over it for at least another 50 to 100 million years. Then a new round of tectonic upheaval began to roil central Arizona. Geologists don't know exactly when this wave of uplift reached Tonto Bridge, but by 65 million years ago, the area had been incorporated into yet another mountain range. Erosion once again began to gnaw at the overlying rocks, exposing the resistant Mazatzal quartzite on the surface for at least the third time since its deposition 1,700 million years ago. Once again it resisted erosion better than the surrounding rocks, and the remnants of the old island were again exhumed as erosion ate away at the more pliant limestone to the east.

This time, the quartzite did not take the shape of an island, but rather a band of bold cliffs flanking a river valley carved in the top of the adjacent limestone. This river deposited the gravels you drove past on your way out of the canyon. As time passed, the accumulation of river gravels built the valley floor up until it reached the top of the Mazatzal cliff, once again leveling out the landscape. This was the geography here 12 million years ago when the mesa-capping basalt flow erupted across the nearly horizontal landscape.

Not long after the eruption, erosion regained the upper hand, and Pine Creek began to carve into the landscape. It breached the basalt cap and soon encountered the nearly indestructible quartzite. Rather than attacking the quartzite head-on, it slid down the flank of the 380-million-year-old island, chipping away at the edges of the softer limestone. This differential erosion has reexposed the Mazatzal quartzite, for the fourth time, in our modern landscape. The fact that the quartzite forms the highest ridge in the immediate vicinity shows that history is repeating itself; this quartzite will continue to stand tall while the entire landscape

Pine Creek has cut its gorge directly along the contact between the resistant Mazatzal quartzite of the right (west) wall of the gorge, and the softer Paleozoic limestone of the left (east) wall.

around it slowly wears away, setting the scene for the next cycle of burial and exhumation in the Mazatzal's long history.

Tonto Bridge State Park is not the only place to see an ancient island in the area. You can actually stand on one at stop 2, the peaceful Christopher Creek Campground east of Payson. The campground lies along the banks of Christopher Creek, which flows below bulky Christopher Mountain, another ancient island of Mazatzal quartzite. As at Tonto Bridge, the creek exhumed the old island by carving a canyon along its flank, at its contact with younger sedimentary rocks. This island stood about 500 feet tall and was comparable in size to the one you observed at the state park.

The sediments comprising the quartzite at Christopher Creek were laid down somewhat later than those exposed at Tonto Bridge. Higher in the Mazatzal depositional sequence, they are dominated by metamorphosed sandstone instead of conglomerate. From campsite 3 on the "A" loop, if you take a fisherman's trail left or east up the canyon about 100 yards, you will see many quartzite slabs marvelously crisscrossed with

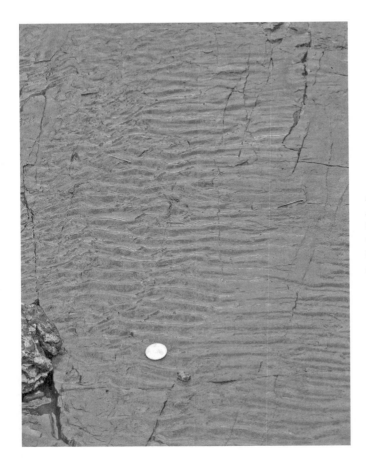

Ripple marks in Mazatzal quartzite upstream from the Christopher Creek Campground; quarter for scale

ripple marks of every imaginable size, pointing in many directions. These were deposited in a large river delta on the edge of a shallow sea in the intra-volcanic-arc basin in which the Mazatzal material accumulated. The ripple marks attest to the complex currents washing over the delta, driven by river flow, tides, waves, and storm surges. The lack of large particles here indicates that by this time the mountains that supplied sediment to the delta, while still sizable, were far less vigorous than they had been when they shed the boulders we found at Tonto Bridge.

Next, head for the campground's picnic area. At the rightmost spot of the right-hand parking lot, immediately beyond the turn off the access road, you are standing at the contact where Paleozoic limestones lapped onto the Christopher Mountain quartzite island over 300 million years ago. Walk 20 yards to the right from the parking area into the small wash there. Tilted layers of maroon to purple Mazatzal quartzite form both the bed of the wash and the crest of the low hill to the northwest. On the near side, horizontal beds of gray limestone cap the rim of the wash. Unlike at Tonto Bridge, where the erosive activity of Pine Creek has erased the contact between the quartzite and the limestone, here you can put your finger on the very place where the sea, represented by the limestone, met the island of quartzite. You are touching a gap in time spanning 1,400 million years—nearly one-third of the earth's history!

At the first two stops, we have seen evidence of the Mazatzal's exposure as an island or hill at 300 million and 13 million years ago, as well as today. Proceed to stop 3 to observe evidence of the rock unit's first exposure 1,200 million years ago. From the group campsite 3 parking lot, walk down the trail behind the ramada to an abandoned road that ends about 300 yards west. Horizontal outcrops of Paleozoic limestone, the same limestone you saw at stops 1 and 2, line the road to its end at a small clearing. In the clearing, very different-looking rocks greet you: outcrops

Rightward-tilted quartzite directly overlain by horizontal layers of Paleozoic limestone on the ancient island shoreline at the Christopher Creek Campground picnic area

of orange and tan siltstone that tilt gently down to the northeast. Many even better outcrops of these rocks lie below along the banks of Sharp Creek, a short but steep cross-country walk, if you are inclined to view them. These rocks are the northernmost outcrops of the Apache group, a sedimentary rock sequence deposited in a shallow sea that stretched across central Arizona 1,200 to 1,000 million years ago.

As Arizona's first sedimentary record after the deposition of the Mazatzal quartzite, the Apache group holds clues about what happened after the Mazatzal Orogeny 1,600 million years ago. From here you can see that the Apache group lies far below the crest of Christopher Mountain, to the south. Like the limestone you saw at stop 2, Apache group rocks lap onto the Mazatzal quartzite, showing that it formed an island here on the edge of the Apache sea. Later, the sediment deposited in that sea was buried, turned to rock, tilted, and reexposed on the flanks of the Christopher Mountain island as rocks of the Apache group, where the horizontal limestone you passed on the walk down from your car was deposited around it. Christopher Mountain, with its apron of Apache group sediments, and similar scattered Mazatzal outcrops to the south provide

Modestly tilted layers of younger Precambrian Apache group rocks in the bed of Sharp Creek

definitive evidence that the Mazatzal quartzite was already incredibly tough and resistant 1,200 million years ago.

The Mazatzal quartzite has, ever since, remained a durable rock unit that has withstood the worst that erosion could throw at it. Again and again, over time, areas veneered by the quartzite protruded above the local topography, islands disrupting the lateral continuity of sedimentary layers around them. These truncations allow us to reconstruct a recurring archipelago of small, craggy islets. Buried by sediments from one sea after another, these pint-sized islands have proven to be some of the most durable aspects of Arizona's geography. They are truly islands in time.

Glossary

agate. A microscopically crystalline form of quartz that displays multiple bands of color.

alluvial fan. A fan-shaped pile of sediment that forms at the base of a mountain range where swiftly flowing streams slow as they reach the flatter valley floor.

andesite. Fine-grained, gray volcanic rock that contains more silica than basalt and less silica than rhyolite.

angular unconformity. A gap in the rock record in which the layers above and below are not parallel.

anticline. An archlike fold in a rock layer.

aquifer. An earth material whose pore spaces contain enough water to be pumped for human needs.

ash. Fine-grained material erupted by volcanoes.

ash fall. A deposit derived from a rain of airborne, fine-grained volcanic ash.

ash flow. A deposit derived from a mixture of highly heated, coarse-grained ash and volcanic gases that races down the flanks of a volcano and along low valleys.

asteroid. A rocky or metallic object in the solar system that is much smaller than a planet.

asthenosphere. A layer in the mantle, below the lithosphere, that is hot enough to flow plastically. This flow drives tectonic plate movement.

badlands. Soft, heavily rilled hills formed of easily eroded mudstone bedrock that are nearly devoid of vegetation.

banded iron formation. A rock consisting of very thin, alternating red (jasper) and steel-gray (iron oxide minerals, e.g., hematite) layers that formed when earth's atmosphere contained almost no oxygen.

basalt. Fine-grained, dark-colored volcanic rock rich in iron-bearing minerals.

Basin and Range. A physiographic region in western North America characterized by parallel mountain ranges separated by valleys. Created by crustal extension.

bedding. Layering of sedimentary rocks.

bedrock. Solid rock underlying superficial loose sediments.

bench. A horizontal rock surface formed on top of a resistant layer or by erosion.

bentonite. A light-colored clay formed from the chemical alteration of volcanic ash that, when wet, expands to over seven times its original volume.

BIF. *See* banded iron formation.

black smoker. A type of submarine hot spring that produces black "smoke" when dissolved metals in the extremely hot water touch cold seawater and crystallize.

bolide. An asteroid or comet that impacts earth from outer space.

bounding surface. A nearly horizontal bedding plane separating sets of crossbeds.

brachiopod. A double-shelled marine invertebrate commonly found fossilized in Paleozoic limestone.

breccia. A sedimentary rock composed of angular, pebble- to boulder-sized fragments in a matrix of finer-grained material.

bryozoan. A marine invertebrate commonly found fossilized in Paleozoic limestone. One type looks like mosquito netting; another, like branched twigs.

butte. An elevated, flat-topped landscape feature with steep to vertical sides that is about as tall as it is wide.

calcite. A light-colored mineral composed of calcium carbonate ($CaCO_3$) that forms limestone.

caldera. A large, basin-shaped volcanic depression, usually formed by an exceptionally violent volcanic eruption.

caprock. A resistant rock layer that overlies softer layers, protecting them from erosion.

carbonate. The compound of carbon and oxygen (CO_3).

cavitation. Implosion of air bubbles in agitated water.

cement (n). Chemically precipitated minerals that bind sediment grains together to form a sedimentary rock.

chemical weathering. Degradation of rocks by chemical processes such as dissolution.

chert. An exceptionally hard sedimentary rock composed of microscopic quartz crystals.

chlorite. A green mica mineral commonly formed during metamorphism of basalt and some other rock types.

cinder. A glassy volcanic rock fragment that falls around a volcanic vent during an eruption.

cinder cone. A conical hill formed of accumulated cinders.

Cladophora. Photosynthetic algae that has formed the base of the aquatic food chain in the Grand Canyon since the damming of the Colorado River.

clast. A particle in a pile of loose sediment or sedimentary rock.

clay. A mineral formed by chemical weathering of preexisting rocks.

cobble. An approximately fist-sized particle in conglomerate or breccia.

Coelophysis. One of the earliest carnivorous dinosaurs, from Triassic time.

coesite. A form of quartz produced only at extremely high pressures; one indicator of possible bolide impact.

collision (zone). Convergence of two tectonic plates composed of continental crust. Large mountain ranges are formed at collision zones.

Colorado Plateau. A physiographic region in western North America characterized by high-elevation plateaus formed from mostly flat-lying sedimentary rocks.

conglomerate. Sedimentary rock that is composed of rounded pebble- to boulder-sized particles in a finer-grained matrix.

continental crust. Buoyant crust composed of silica-rich rocks; forms the continents.

continental shelf. The portion of the shallow ocean floor adjacent to a continent. Made of continental crust.

coprolite(s). Petrified fecal matter.

crinoid. A stalked marine animal commonly found as small, disk-shaped fossils in Paleozoic limestone.

crossbed. A layer of sedimentary rock deposited at an angle to the horizontal. Crossbeds usually mark advancing crests of wind-blown dunes or water-transported sediment.

cross-set. A package of parallel crossbeds.

crust. The outermost shell of the earth, composed of low-density, silica-rich rocks.

cyanobacteria. A type of simple bacteria that photosynthesizes.

dacite. Volcanic igneous rock that contains 63 to 70 percent silica.

delta. An accumulation of sediment that forms where a river meets a standing body of water, such as a lake or ocean.

dendritic. Branching, similar to the vein pattern of a leaf.

differential pressure. Pressure applied to rock that is greater in one direction than in others; characteristically exerted in subduction and collision zones.

dike. A long, narrow igneous intrusion that cuts through massive non-layered rocks or through layered rocks at a high angle.

Dilophosaurus. A large, carnivorous dinosaur with a double crest on its head that lived during early Jurassic time.

dragline. A huge scoop shovel used to mine coal.

eon. The largest division of geologic time. An eon embraces several eras.

era. A division of geologic time. Eras are composed of periods; eons are composed of eras.

erosion. The loosening and moving of soil and rock downhill or downwind. Erosion is accomplished by several agents through a variety of processes.

escarpment. A long, continuous cliff or steep slope that faces in one direction.

Eubrontes giganteus. A distinctive three-toed footprint preserved in sedimentary rocks from early Jurassic time.

evaporite(s). Sedimentary rock composed mainly of minerals left behind when sea or lake water evaporates. Also refers to the minerals themselves.

extension. Pulling apart of the earth's crust, usually along normal faults.

fabric. A recognizable orientation in the crystals of a rock.

fanglomerate. A sedimentary deposit consisting of conglomerate and/or breccia deposited on a steep alluvial fan.

fault. A break in rocks along which movement has occurred.

feldspar. The most common rock-forming mineral; composed of aluminum, silicon, oxygen, and either potassium, sodium, or calcium.

floodplain. A level plain of stratified, unconsolidated sediment on either side of a stream, submerged during floods.

fold. A bent or warped sequence of originally horizontal rock layers.

foliation. The parallel alignment of minerals in metamorphosed rock; occurs when metamorphosing rock is subjected to differential pressure.

formation. The basic unit for the naming of local or regional rock layers.

fossil. A remnant or trace of an organism preserved from the geologic past.

gabbro. A dark-colored igneous rock with the same chemical composition as basalt but with coarser crystals visible to the naked eye.

genus. A taxonomic grouping of related organisms that embraces several species.

geologic time scale. The division of geologic history into eons, eras, periods, and epochs.

granite. A light-colored, coarse-grained intrusive igneous rock with quartz, feldspar, and mica as dominant minerals.

groundwater table. The upper surface of water-saturated rock or sediment beneath the ground.

gypsum. A calcium sulfate mineral that forms when an ocean or lake evaporates.

halite. A sodium chloride mineral formed by evaporation of ocean or lake water. Also known as common table salt.

head frame. A tall, derricklike structure built above a mineshaft. A winch on the head frame raises and lowers miners, materials, and ore in and out of the mine.

head ward erosion. The process of a river extending its length through accelerated erosion along its steep headwaters.

hematite. An iron oxide mineral that is red when its crystals are tiny and steel-gray when they are larger. Forms when primary iron-bearing minerals in a rock are oxidized by exposure to air and water. The mineral that forms common rust.

horn coral. A marine invertebrate commonly found as cornucopia-shaped fossils in Paleozoic limestone.

igneous rock. Rock formed by the solidification of magma.

inclusion. A fragment of older rock within an igneous rock.

interdune area. The land surface between adjacent sand dunes.

intra-volcanic-arc basin. A basin that forms in the middle of a volcanic arc in response to extensional forces.

invertebrate. An animal without a backbone.

ion. An electrically charged atom or molecule.

iridium. A chemical element that is rare on earth and more common in comets and asteroids. Elevated iridium levels in rock may indicate bolide impact.

island arc. A chain of volcanic islands formed at a subduction zone.

isoclinal fold. An exceptionally tight fold in which the two limbs are squeezed so tightly that they are both vertical.

jasper. A type of chert that is colored red due to trace amounts of iron.

joint. A surface of a fracture in a rock along which no displacement has occurred.

knick point. A short, unusually steep reach of river.

Laramide Orogeny. A mountain-building episode 70 to 40 million years ago that raised the Colorado Plateau and southern Rocky Mountains.

lava. Molten material that erupts on the earth's surface.

lava dam. A river blockage created by the flow of lava into a river canyon.

law of lateral continuity. One of Nicolas Steno's three laws of sedimentary geology. It states that sedimentary layers are deposited in extensive, continuous sheets.

law of original horizontality. One of Nicolas Steno's three laws of sedimentary geology. It states that sedimentary layers are deposited horizontally.

law of superposition. One of Nicolas Steno's three laws of sedimentary geology. It states that, in an undisturbed stack of sedimentary layers, the one on the bottom is the oldest and the one on the top is the youngest.

limestone. A sedimentary rock composed mostly of the mineral calcite and often containing marine fossils.

listric fault. A fault that is steep near the surface but flattens out with depth. This change in angle causes rock layers that move along the fault to rotate and, therefore, tilt.

lithification. The compaction and cementation of loose sediments into sedimentary rock.

lobe. A curved or rounded projection or division.

maar. A broad volcanic crater formed by one or more explosive phreatic eruptions.

magma. Molten material below the earth's surface, from which igneous rock is derived.

magma chamber. A pool of molten rock beneath the earth's surface.

magnetite. A magnetic iron oxide mineral.

mantle. The section of the earth's interior between the crust and the core.

massive sulfide. A deposit of sulfide minerals without distinct layering. Commonly forms an ore for copper, gold, and other precious metals.

Massospondylus. A plant-eating dinosaur during early Jurassic time.

matrix. An aggregation of small crystals or particles that fills in the spaces between larger crystals in igneous rock and between gravel particles in sedimentary rock.

Mazatzal Orogeny. A mountain-building episode 1,600 million years ago that deformed rock layers in central Arizona.

meander. A broad, semicircular curve in a stream that develops from erosion on the outside of a bend and deposition on the inside.

member. A subdivision of a rock formation that possesses distinctive characteristics.

mesa. An elevated, steep-sided, flat-topped landscape feature that is much wider than it is tall.

metamorphic rock. Rock whose original mineralogy, texture, or composition has been changed by the effects of pressure and/or temperature.

meteorite. A space rock that has landed on the earth's surface.

metoposaur. A large carnivorous amphibian that lived during Triassic time.

mica. A group of silicate minerals that are distinctive because they form in two-dimensional sheets, like the pages of a book.

microcontinent. A mass of continental crust that is smaller than a continent but larger than an island.

mid-ocean ridge. A submarine mountain chain on the sea floor that forms where two tectonic plates are spreading apart.

mineral. A naturally occurring, inorganic crystalline solid with a specific chemical composition.

mollusk. An invertebrate animal with a bilaterally symmetrical body.

monocline. A steep, one-armed fold in otherwise horizontal layers of rock, similar to a riser between treads of a staircase.

mudstone. Fine-grained sedimentary rock formed from hardened clay and silt.

natural bridge. An archlike rock formation carved by erosion that spans a river or other watercourse.

nodule. A lump of one rock type (often chert) encased within another rock type.

normal fault. A fault due to crustal extension in which the block resting on top of the angling fault plane slides down relative to the block below.

oceanic crust. Dense crust composed of basalt that forms ocean basins.

opaline silica. A type of quartz crystal that incorporates water into its crystalline structure. Produced by marine plankton and during weathering of volcanic rocks.

open pit mine. A mine in which ore is extracted by excavation of a large pit rather than through underground shafts.

ore. A rock unit in which a valuable material, e.g., copper, is concentrated to such an extent that it is profitable to mine it.

orogeny. A tectonic episode of mountain-building.

outcrop. A segment of exposed bedrock on the earth's surface.

paleocanyon. An ancient canyon.

paleogeography. The reconstruction of a region's past geography.

paleontology. The study of fossil organisms and traces (such as tracks) of prehistoric organisms in the rock record.

Pangaea. The supercontinent between 300 and 200 million years ago that consisted of virtually all of the earth's landmass.

perched groundwater table. A local groundwater table situated above the regional groundwater table due to the presence of a higher impermeable layer.

perennial lake. A lake filled with water year-round.

period. The fundamental unit of geologic time; a subdivision of an era.

permineralization. A process of fossilization in which mineral materials fill the pore spaces between hard portions of a dead organism.

petrification. Conversion of organic material into rock.

phreatic eruption. An explosive type of volcanic eruption in which underground magma flashes water to steam.

phyllite. A metamorphic rock consisting of greater than 50 percent mica minerals. Resembles schist, but the individual mica crystals are smaller.

physical weathering. Degradation of rocks by physical processes such as abrasion.

pillow basalt. A type of basalt with a texture of round, pillowlike structures. Forms when a basalt lava flow solidifies under water.

plateau. An elevated, flat-topped landscape feature.

plate tectonics. A theory that earth's crust is broken into separate plates that move and interact with each other.

playa. A desert valley floor where an ephemeral lake periodically dries out, leaving behind evaporite minerals and fine-grained sediments.

porous. Possessing a high proportion of pore spaces relative to solid mineral grains.

Postosuchus. A large, carnivorous amphibian of Triassic time.

precipitate. To solidify out from a dissolved state, e.g., an ion coming out of solution. As a noun, refers to material that has precipitated.

pumice. A glassy volcanic material erupted from a volcano.

pyrite. An iron sulfide mineral, also known as fool's gold.

quartz. A very hard, clear, or translucent mineral composed of silica (SiO_2).

quartzite. A metamorphic rock composed of sand-sized quartz grains fused by heat and pressure.

recharge. The process by which surface water percolates into the ground and becomes groundwater.

recharge area. A region on the earth's surface where water percolates into the ground to feed an aquifer.

rectangular drainage. A series of streams that flow in a rectangular path, with many right angles. Commonly forms in young landscapes where drainages are controlled by joints.

reverse fault. A compressional fault in which the block above (resting on) the fault plane moves up relative to the block below.

rhyolite. A light-colored volcanic rock type that contains over 70 percent silica. Chemically identical to granite, but with smaller crystals.

ring fracture. A circular fault that forms around the edge of a caldera volcano as it collapses into the evacuated magma chamber.

ripple mark. A wavy mark on the bedding surface of sedimentary rock formed by moving water or wind.

salt tectonics. The deformation of rock layers due to the flow of evaporites, or salts. Evaporites flow upward because they are less dense than surrounding minerals.

sandstone. Sedimentary rock composed of cemented sand-sized grains.

sanidine. A variety of potassium feldspar mineral found in volcanic rocks.

Sanmiguelia lewisii. A fossil plant believed to be either the first flowering plant or its direct ancestor.

scarp retreat. Gradual retreat of a cliff band caused either by spring sapping or by undercutting by a river.

schist. Foliated metamorphic rock composed of plate-shaped mica minerals aligned in the same direction.

sedimentary rock. Rock formed on the earth's surface from the weathered products of preexisting rocks or as the product of biological and/or chemical processes.

Sevier-Mogollon Highlands. A former mountain range uplifted by the Sevier and Laramide Orogenies in the area of the modern Basin and Range province of western and southern Arizona.

Sevier Orogeny. A Cretaceous mountain-building episode that uplifted the region now occupied by the Basin and Range province.

shale. Fine-grained sedimentary rock formed from hardened clay and silt; typically splits into thin layers.

silica. A tetrahedral compound of one silicon atom surrounded by four oxygen atoms that forms the basic building block of most rock-forming minerals.

silt. Sediment consisting of particles finer than sand and coarser than clay.

sinkhole. A depression created by the collapse of a cave roof.

smelter. An industrial plant that extracts metals from mined ores through heating.

species. A group of similar organisms that can interbreed and produce fertile offspring.

spherule. A small blob of melted rock ejected and melted by a bolide impact.

spicule. A tiny, silica-rich support structure that stiffens the tissues of sponges.

sponge. A marine invertebrate commonly found as brainlike fossils in Paleozoic limestone.

spring sapping. Erosion of a cliff face by springwater issuing from its base.

stalactite. A conical calcite deposit hanging from the ceiling of a cave, precipitated by dripping water.

strata. Layers of sedimentary rock.

stream piracy. Process by which the headwaters of a steeper stream erode headward, intercept a lower-gradient stream, and "capture" that water into their steeper stream course.

stretch-pebble conglomerate. A conglomerate so metamorphosed and sheared that individual pebbles are severely stretched.

subduction. A tectonic process in which a dense oceanic plate dives beneath another plate due to convergence.

subduction zone. The area on a plate boundary where subduction occurs. Consists of an oceanic trench and a chain of volcanoes.

supercontinent. An ancient continent that contained all or most of the world's existing continental area.

swallet. An underground outlet for a lake.

tailings. Ground-up waste rock produced during the extraction of metals from ores.

talus. Large, angular rock fragments found at the base of a cliff or other steep slope from which they have eroded.

tectonic. Pertaining to the large-scale processes that deform the earth's crust. Mostly used in reference to plate tectonics.

tidal channel. A major channel through which tidal currents move, extending from offshore into a tidal marsh or flat.

tidal flat. A very flat mud- or sand-covered coastal plain alternately exposed and covered by sea tides.

tidewater glacier. A glacier that empties into the sea.

track. A fossilized footprint.

trackway. A series of fossilized footprints made by a moving animal.

trade winds. A wind pattern in tropical latitudes that blows consistently from the east.

travertine. A form of limestone deposited by springs or in caves.

trench, ocean. A long, narrow depression in the sea floor that develops where one tectonic plate dives beneath another in a subduction zone.

tridactyl print. A footprint consisting of three toes.

tufa. A soft form of limestone deposited by a spring. Travertine is its hard, dense equivalent.

tuff. Rock composed of compacted volcanic fragments.

tuff ring. A ring of tuff that surrounds the crater from which it was ejected during a volcanic eruption.

unconformity. A surface that represents a time gap in the geologic record where rock layers were eroded or never deposited.

underprint. An impression left in sediment layers up to several inches below the surface.

vesicle. A small cavity in a volcanic rock created by a trapped gas bubble.

viscosity. A material's internal resistance to flow. The higher the viscosity, the slower the flow.

volcanic (island) arc. A chain of volcanoes associated with a subduction zone. When the plates involved in subduction are oceanic, the volcanoes form an arcing chain of islands. When one plate is continental, the volcanoes form a coastal mountain range.

volcanic neck. A tower or butte formed of the solidified magma that filled the conduit of an extinct volcano.

volcanic rock. Igneous rock that cooled on the earth's surface; characterized by crystals too small to be seen with the naked eye.

weathering. Disintegration and decomposition of rock at or near the earth's surface. See also **chemical weathering**; **physical weathering**.

welded tuff. A volcanic rock formed from ash welded into a coherent, reasonably strong mass. It forms near the base of volcanic ash flows, where ash is thick enough to provide the necessary pressure for welding.

white smoker. A submarine hot spring from which dissolved silica precipitates upon contact with cold seawater, appearing as white smoke.

Yavapai Arc. A complex group of volcanic island arcs off the coast of the proto–North American continent that progressively collided with the continent about 1,700 million years ago.

Sources of More Information

General

Abbott, Lon, and Terri Cook. 2004. *Hiking the Grand Canyon's Geology*. Seattle: Mountaineers Books.

Baldridge, W. Scott. 2004. *Geology of the American Southwest*. Cambridge, UK: Cambridge University Press.

Lucchitta, Ivo. 2001. *Hiking Arizona's Geology*. Seattle: Mountaineers Books.

Nations, Dale, and Edmund Stump. 1996. *Geology of Arizona*. Dubuque, IA: Kendall/Hunt Publishing Co.

1. Pearce Ferry

Lucchitta, Ivo. 1987. The Mouth of the Grand Canyon and Edge of the Colorado Plateau in the Upper Lake Mead Area, Arizona. In *Geological Society of America Centennial Field Guide*, vol. 2, ed. Stanley Beus, 365–70. Boulder, CO: Geological Society of America.

Ranney, Wayne. 2005. *Carving Grand Canyon*. Grand Canyon, AZ: Grand Canyon Association.

Young, Richard, and Earle Spamer, eds. 2001. *Colorado River: Origin and Evolution*. Grand Canyon, AZ: Grand Canyon Association. (Chapters 12–14 and 32 are particularly relevant.)

2. Toroweap Overlook

Abbott, Lon, and Terri Cook. 2004. *Hiking the Grand Canyon's Geology*. Seattle: Mountaineers Books, 262–70.

Billingsley, George. 2001. Volcanic Rocks of the Grand Canyon Area. In *Colorado River: Origin and Evolution*, ed. Richard Young and Earle Spamer, chap. 33. Grand Canyon, AZ: Grand Canyon Association.

Hamblin, W. K. 2003. Late Cenozoic Lava Dams in the Western Grand Canyon. In *Grand Canyon Geology*, ed. Stanley Beus and Michael Morales, chap. 17. New York: Oxford University Press.

Ranney, Wayne. 2005. *Carving Grand Canyon*. Grand Canyon, AZ: Grand Canyon Association.

Spencer, Jon, and Philip Pearthree. 2005. Abrupt Initiation of the Colorado River and Initial Incision of the Grand Canyon. *Arizona Geology* 35(4):1–4.

3. The Colorado River and Glen Canyon Dam

Grand Canyon Monitoring and Research Center website, www.gcmrc.gov.

Lucchitta, Ivo, and Luna Leopold. 1999. Floods and Sandbars in the Grand Canyon. *GSA Today* 9:1–7.

Rubin, David, and others. 2002. Recent Sediment Studies Refute Glen Canyon Dam Hypothesis. *Eos* 83:273, 277–78.

Shannon, Joseph, and Emma Benenati. *Essentials of Aquatic Ecology in the Colorado River.* Flagstaff, AZ: NAU Creative Communications.

4. Antelope Canyon

Chan, Marjorie, and Allen Archer. 2000. Cyclic Eolian Stratification on the Jurassic Navajo Sandstone, Zion National Park: Periodicities and Implications for Paleoclimate. In *Geology of Utah's Parks and Monuments*, ed. D. A. Sprinkel, T. C. Chidsey, and B. P. Anderson, 607–17. Publication 28. Salt Lake City: Utah Geological Association.

Loope, D. B., and C. M. Rowe. 2003. Long-Lived Pluvial Episodes during Deposition of the Navajo Sandstone. *Journal of Geology* 111:223–32.

Navajo Parks and Recreation Department website, www.navajonationparks.org.

5. Navajo National Monument and Black Mesa

Chronic, Halka, and Lucy Chronic. 2004. *Pages of Stone: Geology of the Grand Canyon and Plateau Country National Parks and Monuments.* Seattle: Mountaineers Books, 127–30.

Nations, Dale, and Edmund Stump. 1996. *Geology of Arizona.* Dubuque, IA: Kendall/Hunt Publishing Co. (Chapters 11 and 15 are particularly relevant.)

Zhu, Chen, and others. 1998. Responses of Ground Water in the Black Mesa Basin, Northeastern Arizona, to Paleoclimatic Changes during the Late Pleistocene and Holocene. *Geology* 26:127–30.

6. Monument Valley

Baars, Donald. 1995. *Navajo Country.* Albuquerque: University of New Mexico Press.

Baldridge, W. Scott. 2004. *Geology of the American Southwest.* Cambridge, UK: Cambridge University Press.

Chenoweth, William. 2000. Geology of Monument Valley Navajo Tribal Park, Utah-Arizona. In *Geology of Utah's Parks and Monuments*, ed. D. A. Sprinkel, T. C. Chidsey, and P. B. Anderson, 529–33. Publication 28. Salt Lake City: Utah Geological Association.

Hopkins, R. 2002. *Hiking the Southwest's Geology: Four Corners Region.* Seattle: Mountaineers Books, 109–12.

7. The Grand Canyon and the Gorge of the Little Colorado River

Abbott, Lon, and Terri Cook. 2004. *Hiking the Grand Canyon's Geology.* Seattle: Mountaineers Books.

Beus, Stanley, and Michael Morales. 2003. *Grand Canyon Geology.* New York: Oxford University Press. (Chapters 9–12 are particularly relevant.)

Mathis, Allyson, and Carl Bowman. 2005. What's in a Number? Numeric Ages for Rocks Exposed within Grand Canyon, Part 2. *Nature Notes* 21 (2):1–5.

Price, L. Greer. 1999. *An Introduction to Grand Canyon Geology.* Grand Canyon, AZ: Grand Canyon Association.

Ranney, Wayne. 2005. *Carving Grand Canyon.* Grand Canyon, AZ: Grand Canyon Association.

Sadler, Christa. 2005. *Life in Stone.* Grand Canyon, AZ: Grand Canyon Association.

8. Moenave Dinosaur Tracks

DeCourten, F. 1998. *Dinosaurs of Utah.* Salt Lake City: University of Utah Press, 66–85.

Lockley, M. 1995. *Dinosaur Tracks and Other Fossil Footprints of the Western United States.* New York: Columbia University Press, 109–22.

Morales, M. 1986. Dinosaur Tracks in the Lower Jurassic Kayenta Formation near Tuba City, Arizona. In *Dinosaur Tracksites,* ed. M. Lockley, 14–15. *University of Colorado at Denver Geology Department Magazine,* special issue 1.

Olsen, P. E., and others. 2002. Ascent of Dinosaurs Linked to an Iridium Anomaly at the Triassic-Jurassic Boundary. *Science* 296:1,305–7.

Welles, S. P. 1971. Dinosaur Footprints from the Kayenta Formation of Northern Arizona. *Plateau* 44:27–38.

9. Canyon de Chelly National Monument

Baars, Donald. 1995. *Navajo Country.* Albuquerque: University of New Mexico Press.

Chronic, Halka, and Lucy Chronic. 2004. *Pages of Stone: Geology of the Grand Canyon and Plateau Country National Parks and Monuments.* Seattle: Mountaineers Books, 60–64.

Lucchitta, Ivo. 2001. *Hiking Arizona's Geology.* Seattle: Mountaineers Books. Hike 10: White House Ruin Trail, 111–13.

10. The Peach Springs Tuff

Hillhouse, J. W., and R. E. Wells. 1991. Magnetic Fabric, Flow Directions, and Source Area of the Lower Miocene Peach Springs Tuff in Arizona, California, and Nevada. *Journal of Geophysical Research* 96:12,443–60.

Nielson, J. E., and others. 1990. Age of the Peach Springs Tuff, Southern California and Western Arizona. *Journal of Geophysical Research* 95:571–80.

Wells, R. E., and J. W. Hillhouse. 1989. Paleomagnetism and Tectonic Rotation of the Lower Miocene Peach Springs Tuff: Colorado Plateau, Arizona to Barstow, California. *Geological Society of America Bulletin* 101:846–63.

Young, R. A., and others. 1987. Geomorphology and Structure of the Colorado Plateau/Basin and Range Transition Zone, Arizona. In *Geologic Diversity of Arizona and Its Margins: Excursions to Choice Areas*, ed. G. E. Davis and E. M. VandenDolder, 182–86. Geological Survey Branch Special Paper 5. Tucson: Arizona Bureau of Geology and Mineral Technology.

11. The San Francisco Volcanic Field

Duffield, W. 1997. *Volcanoes of Northern Arizona.* Grand Canyon, AZ: Grand Canyon Association.

Holm, Richard. 2004. Landslide Preconditions and Collapse of the San Francisco Mountain Composite Volcano, Arizona, into Cold Debris Avalanches in Late Pleistocene. *Journal of Geology* 112:335–48.

Holm, Richard, and George Ulrich. 1987. Late Cenozoic Volcanism in the San Francisco and Mormon Volcanic Fields, Southern Colorado Plateau, Arizona. In *Geologic Diversity of Arizona and Its Margins: Excursions to Choice Areas*, ed. G. E. Davis and E. M. VandenDolder, 85–94. Geological Survey Branch Special Paper 5. Tucson: Arizona Bureau of Geology and Mineral Technology.

Lucchitta, Ivo. 2001. *Hiking Arizona's Geology.* Seattle: Mountaineers Books, 114–34.

Nations, Dale, and Edmund Stump. 1996. *Geology of Arizona.* Dubuque, IA: Kendall/Hunt Publishing Co., 181–201.

12. Grand Falls

Bezy, J. V. 2003. Stream Displaced by a Lava Flow: Grand Falls. In *Guide to the Geology of the Flagstaff Area*. Tucson: Arizona Geological Survey, 44–45.

Duffield, Wendell, and others. 2006. Multiple Constraints on the Age of a Pleistocene Lava Dam across the Little Colorado River at Grand Falls, Arizona. *Geological Society of America Bulletin* 118:421–29.

Hopkins, R. 2002. *Hiking the Southwest's Geology: Four Corners Region*. Seattle: Mountaineers Books, 67–71.

13. The Sedona Red Rocks

Blakey, Ronald, and Larry Middleton. 1987. Late Paleozoic Depositional Systems, Sedona-Jerome Area, Central Arizona. In *Geologic Diversity of Arizona and Its Margins: Excursions to Choice Areas*, ed. G. E. Davis and E. M. VandenDolder, 143–57. Geological Survey Branch Special Paper 5. Tucson: Arizona Bureau of Geology and Mineral Technology.

Ranney, Wayne. 2001. *Sedona through Time*. Flagstaff, AZ: Zia Interpretive Services.

14. Meteor Crater

Melosh, H., and G. Collins. 2005. Meteor Crater Formed by Low-Velocity Impact. *Nature* 434:157.

Meteor Crater website, www.barringercrater.com/science/main.htm.

Nations, Dale, and Edmund Stump. 1996. *Geology of Arizona*. Dubuque, IA: Kendall/Hunt Publishing Co., 203–11.

Shoemaker, E. M. 1987. Meteor Crater, Arizona. In *Rocky Mountain Centennial Field Guide*, vol. 2, ed. S. Beus, 399–405. Boulder, CO: Geological Society of America.

Shoemaker, E. M., and S. Kieffer. 1974. *Guidebook to the Geology of Meteor Crater, Arizona*. Tempe: Arizona State University Center for Meteorite Studies.

15. Petrified Forest National Park

Bezy, J., and A. Tevena. 2000. *Guide to Geologic Features at Petrified Forest National Park*. Down-to-Earth series 10. Tucson: Arizona Geological Survey.

Harris, A., E. Tuttle, and S. Tuttle. 2004. *Geology of National Parks*. Dubuque, IA: Kendall/Hunt Publishing Co., 102–12.

Long, R., and R. Houk. 1988. *Dawn of the Dinosaurs*. Petrified Forest, AZ: Petrified Forest Museum Association.

National Park Service website, www2.nature.nps.gov/geology/paleontology/publications/cfm.

16. The Rocks of Prescott

Conway, Clay, and others. 1987. Tectonic and Magmatic Contrasts across a Two-Province Proterozoic Boundary in Central Arizona. In *Geologic Diversity of Arizona and Its Margins: Excursions to Choice Areas*, ed. G. E. Davis and E. M. VandenDolder, 158–75. Geological Survey Branch Special Paper 5. Tucson: Arizona Bureau of Geology and Mineral Technology.

Karlstrom, K., and S. Bowring. 1993. Proterozoic Orogenic History of Arizona. In *Precambrian: Conterminous U.S.*, vol. C-2 of *Geology of North America*, ed. J. Reed and others, 188–211. Boulder, CO: Geological Society of America.

Krieger, Medora. 1965. Geology of the Prescott and Paulden Quadrangles. U.S. Geological Survey Professional Paper 467. U.S. Department of the Interior, Government Printing Office.

Slat, Roger, and others. 1978. Precambrian Pikes Peak Iron-Formation, Central Arizona. In *Guidebook to the Geology of Central Arizona*, ed. Donald Burt and Troy Pewe, 73–82. Special Paper 2. Tucson: Arizona Bureau of Geology and Mineral Technology.

17. The Mines of Jerome

Lindberg, Paul, and Mae Gustin. 1987. Field-Trip Guide to the Geology, Structure, and Alteration of the Jerome, Arizona Ore Deposits. In *Geologic Diversity of Arizona and Its Margins: Excursions to Choice Areas*, ed. G. E. Davis and E. M. VandenDolder, 176–81. Geological Survey Branch Special Paper 5. Tucson: Arizona Bureau of Geology and Mineral Technology.

Ranney, Wayne. 1989. The Verde Valley: A Geological History. *Plateau* 60(3).

18. Montezuma Castle National Monument

Holm, R. F., and others. 1998. Miocene Volcanism and Geomorphology in Verde Valley, and Petrology of Alkaline and Mildly Alkaline Rocks at House Mountain Shield Volcano, Sedona, Arizona. In *Geologic Excursions in Northern and Central Arizona*, ed. E. M. Duebendorfer, 1–16. Flagstaff: Northern Arizona University Department of Geology.

Ranney, Wayne. 1989. The Verde Valley: A Geological History. *Plateau* 60(3).

19. Tonto Bridge State Park

Fellows, Larry. 1990. Tonto Natural Bridge, Arizona's Newest State Park, the First 1.7 Billion Years. *Arizona Geology* 20:8–10.

20. The Mazatzal Quartzite

Baldridge, W. Scott. 2004. *Geology of the American Southwest.* Cambridge, UK: Cambridge University Press.

Cox, Ronadh, and others. 2002. Sedimentology, Stratigraphy, and Geochronology of the Proterozoic Mazatzal Group, Central Arizona. *GSA Bulletin* 114:1,535–49.

Wrucke, C. 1993. The Apache Group, Troy Quartzite, and Diabase: Middle Proterozoic Rocks of Central and Southern Arizona. In *Precambrian: Conterminous U.S.*, vol. C-2 of *Geology of North America*, ed. J. Reed and others, 517–21. Boulder, CO: Geological Society of America.

Index

Numbers in *italics* indicate an illustration.

acid rain, 200, 245
agate, 211
Agathla, 84, *85*
alcoves: at Betatakin, 65, *66*; formation of, 69–70, *70*; at Montezuma Castle, 7, 252–3, *252*; at Tonto Bridge State Park, 265; at White House Ruin, 122, *123*, 133
alluvial fans, *273*, 273–4
Alvarez, Luis and Walter, 199
Anasazi. *See* Ancestral Pueblo culture
Ancestral Pueblo culture (Anasazi), 65, 74, 122, 173, 252
Ancestral Rocky Mountains, 4, 128, *175*, 179, *180*
andesite, *153*, 155–8, *158*, 164, 217. *See also* Sinagua formation
angular unconformity, 13, *14*, 15
Antelope Canyon, 52–63
Apache group, 3–4, 281–82, *282*
Apache (Yavapai) Indians, 253
Appalachian Mountains, 4, *175*
aquifer, 69–70, 76–77. *See also* N-aquifer
arch, natural. *See* natural bridge
Archaea bacteria, 241
ash: and badlands, 208–9; eruption, 157; fall deposits, 136–37; flow deposits, 6, 136–37, 139, 146–47; formation from lava, 135–36, 221; and petrification, 211; and stratovolcanoes, 157
asteroid(s), 191; and end of Cretaceous, 6; and end of Paleozoic, 4; and end of Triassic, 5, 117–18; impacts, frequency of, 198; iridium in, 117, 199; and Meteor Crater, 7. *See also* bolides; Meteor Crater; meteorite(s); meteor(s)
asthenosphere, 217
astronaut training, 152, 191
Augustine, Mount, 220

Babbitt, Bruce, 160
bacteria, 4, 227, 241, 267–68, *268*
badlands, 86, *204*, 203–5, 207–8
banded iron formations (BIFs), 3, *226*, 226–27
Barringer, Daniel, 193
basalt, *153*, 153; and assembly of Rodinia, 3; around Flagstaff, 153; generation of, 164; at Grand Falls, 167–71, *169*, *170*, 174; around Jerome, 237, 239; in lava cascades in Grand Canyon, *25*, 32, 37; at Montezuma Castle, 258–59, *259*; in oceanic crust, 217; (metamorphosed) around Prescott, *228*, 228–30; and silica content, 153, *153*; in Tonto Bridge State Park, 262, 265, *266*, *276*, 278; around Verde Valley, *250*; vesicles in, *155*, 155. *See also* pillow basalt
Basin and Range (province), 9; and Colorado Plateau, 7, 14; and Colorado River, 18–21; formation of, 6–7, 35, 147, *147*, 242; and Grand Canyon, 22; and San Andreas fault, 35, 139, *143*;

Basin and Range (*continued*)
and Verde Valley, 249; and volcanism, *143*, 144, 164. *See also* normal faults; Peach Springs tuff
beaches, Colorado River, 44–47, 49–50, 173
bedding plane(s), 254, 273
bees, ancient, 213
Bell Rock member, 181–83, *182*, *183*, 185
bentonite, 208, *208*
Betatakin, 65–66, *66*
BIF. *See* banded iron formations
Black Mesa, 65–77, 126, *126*
Black Mesa Mine, 76
Blue Mesa Trail, 204–9
bolides, 198–99
Bonito lava flow, 152–55, *155*
bounding surfaces, *55*, 55–56, *56*, *101*, *130*. *See also* cross-sets
brachiopods, 96, *96*, *97*
breccia, *159*, 161, *162*
Bright Angel shale, 29
bryozoans, 96, *97*
Bureau of Reclamation, 46, 49–50
butte(s), 79, 86, 89–90, *91*. *See also* Merrick Butte; Mitten Butte(s); Shinarump conglomerate

calcite, 257–58, *258*
caldera, 239; eruption, 135–36, *138*, 139, 157, 239; and Haigler rhyolite, 274; Jerome, 239–40, *240*, 242. *See also* Ngorongoro Crater; ring fracture
caliche, 179
Canyon de Chelly, 4, 5, 6, 120–33, *121*. *See also* DeChelly sandstone; Defiance uplift; Shinarump conglomerate
Canyon Diablo, 191
carbon, 47, 72, 74, 209
Carson, Kit, 122

casts, limestone, 257
caves, *253*, 254, *255*, 267
cavitation, 58
channels (ancient river), 4, 5, 131, *132*, 178–79, *178*
chert: in BIFs, 226, *226*, 227; depositional environment of, 225; erosion of, 98; nodules, 95–98, 102–3; in petrified trees, 211; precipitation around organic material, 98; in stretch-pebble conglomerate, 225. *See also* agate
Chicxulub crater, 198–200
Chinle, 125
Chinle formation, 81–82; badlands in, 86, *204*, 203–5, 207–8; at Canyon de Chelly, 124, *124*; and climate, 206; color of, 209, 211; in Comb Ridge monocline, *84*; depositional environments of, 81–82, 206; erosion of, 171; fossils in, 203, 206, 213–16; and Monument Valley, 84, 86; in Petrified Forest National Park, 203–11; and Triassic ecosystem, 117, *213*, 212–16. *See also* petrified wood; Shinarump conglomerate; Sonsela member
Christopher Creek, 279; campground, 279–81
Christopher Mountain, 279–80, *280*, *281*
Chuska mountain range, 123
cinder cones, 32–33, *154*, 167. *See also* Merriam Crater; Sunset Crater National Monument
cinders, *154*, 154–55, 167
Cladophora, 47
Clarkdale smelter, 244, *247*
clasts, 272
clay, 208, *208*, 257
Cleopatra formation, 235, 237, 239, 242

Cleopatra Hill, 235, *236*
cliff dwellings, 122. *See also* Betatakin; White House Ruin
climate: and Ancestral Rockies, 179; asteroid impact effects on, 117, 199–200; and Lake Verde, 249; during Mesozoic, 79, 82; and Pangaea, 57, 179–80; during Permian, 100–102, 179; during Pleistocene, 250; and sediment deposition, 100; during Triassic, 206
coal, 6, 73–77. *See also* Black Mesa Mine; Kayenta Mine; Peabody Coal Company
Coconino sandstone: age of, 100; depositional environment of, 4, 100–101; in Grand Canyon, *93, 94*, 95, *101*, 102; lateral changes in, 102, 104; in Little Colorado River gorge, *103*, 104; at Meteor Crater, 196–97; near Sedona, *177*, 181–82, *183, 186*, 187–88; tracks in, 188; transition from Sycamore Pass member, 182
Coelophysis, 116–17, *213, 215*, 215–16
coesite, 196
Colorado Plateau (province): and Basin and Range, 7; boundary of, 7, 9, 14, 141, 261; destruction of, 34–37; erosion of, 22, 63; uplift of, 6, 29–30, 67, 82, 127; and Verde Valley, 249; and Wheeler Ridge, 11, *17*
Colorado River, 39–51; ancestral, 22–24; ancient canyons in, *145*, 146; apportionment of, 40–42; basin, *41*; carving of Grand Canyon by, 29–30, *31*, 95, 98; fish, 48–50; flood deposits, 45, *46*; floods, 45–46, 49–50; and Glen Canyon Dam, 38–39, 44–45, 47–49; Hualapai legend of, 50; and lava cascades, 25, 32; and lava dams, 7, 32–34; as master stream, 9, 63; and Muddy Creek formation, 18–21; sandbars and beaches of, 44–47, 49–50; and sediment, 44–45, 47, 49–50; tributaries, 44, 50, 173. *See also* Little Colorado River; Paria River
Comb Ridge, 83–84; monocline, *83*, 83–84, *84*
comets. *See* bolides
compression, 142, *142*, 230
concretions, 119
conglomerate, 272–74, *272*. *See also* Shinarump conglomerate; stretch-pebble conglomerate
continental crust, 127, 217, *219*; composition of, 164; and northern Arizona, 1; partial melting of, 164; in volcanic arcs, 219
continental shelf, 96
continents, 5, 217, 219. *See also* North America
copper, 233; extraction of, 244–45; minerals, 236; ores, 3, 233, 235; value of, 235. *See also* slag; tailings
coprolites, 119, *119*, 212
coral, horn, 96, *97*
Cottonwood smelter, 244
crater: Meteor, 189–201; volcanic, *154*. *See also* impact crater
Cretaceous Period, 5, 72. *See also* Dakota sandstone; Mancos shale; Mesa Verde group; Sevier-Mogollon Highlands
crinoids, 96, *97*
crossbeds, *55*, 55–56, *56*; at Antelope Canyon, 56; in Canyon de Chelly, 125, *125*, 129, 131; at Grand Canyon, 100, *101*; high-angle vs. low-angle, 129; in Mazatzal quartzite, 274;

crossbeds (*continued*)
  at Navajo National Monument, 65; near Sedona, *185*, 186–7
cross-sets, 55–56, 129, *130*
crust. *See* continental crust; oceanic crust
cyanobacteria, 227

dacite, *153*, 155, 162, 164
Dakota sandstone, 72, *73*
dams: on Colorado River, 8, *41*; lava, 7, 32–34, 168, *169*, 170; on Little Colorado River, 168–70, *169*. *See also individual dams*
dating, tree-ring, 65, 154
debris rafting, 259, *259*
DeChelly sandstone: at Canyon de Chelly, 125, 128–31, *130*, 133; depositional environment of, 4, 79, 129–31, *130*, *186*; fossils in, 130; in Monument Valley, 79, 86–89, *90*; and Shinarump conglomerate, *125*, *132*; tracks in, 130
Defiance: monocline, 123; Plateau, 123–24, 126–28, 131, 133
desert(s): and Coconino sandstone, 100; and DeChelly sandstone, 79, 129–30; largest on North America, 82; and the Navajo sandstone, 42, 54, 57, 65, 82; on Pangaea, 180; Sahara, 54, 82, 101
Desert View overlook, 102, 104–5
differential pressure, 221
Dilophosaurus, 107–18, *113*. *See also Coelophysis*
Diné. *See* Navajo
dinosaur(s), 106–18; claws, 107, 110, *110*, *111*, 112; "egg," 119; evolution of, 5–6, 116–18; extinction of, 116–18, 199–201; feces, 119, *119*; meat-eating, 107, 113–14, 215;

plant-eating, 114. *See also Coelophysis*; *Dilophosaurus*; *Massospondylus*; paleontologists; tracks
Douglas, James, 238
dragline, 74, *75*
drainage networks, 61, *63*
drought, 5, 76
dunes. *See* sand dunes

earth, 1
ecosystems, 65, 74, 75. *See also* Grand Canyon; Triassic Period
endangered species, in Grand Canyon, 45–50, 173
eons, 1. *See also individual eons*
eras, 1, 4–5. *See also individual eras*
erosion: of ancient mountains, 3–4, 274–75; of badlands, 208; of Black Mesa, 69; at Canyon de Chelly, 123; of chert, 98; on Colorado Plateau, 63; by Colorado River, 95, 98; deepening vs. widening, 29, 89, 95; of Defiance Plateau, 126, *126*; differential, 171, 278; and dinosaur tracks, 111; and freeze-thaw cycles, 88–89; in Grand Canyon, 98; at Grand Falls, 170–73, *172*; of granite, 209; headward, 7, 22, *23*; of impact craters, 198; along joints, 61, *62*, 63, 88–90, *90*; of Kaibab formation, 98; of lava dams, 34; of Mazatzal quartzite, 264–65, 276, 278; of the Monument upwarp, *83*, 84, 87; in Monument Valley, 86–91, *90*, *91*; and the Navajo sandstone, 42; in northern Arizona, 3; resistance of quartzite to, 271; river, 7, 58–59, 95, 171–73, *172*; of San Francisco Mountain,

163; of sedimentary rocks, 29, 67, 89, 95, 171; of Shonto Plateau, *68*, 69; of volcanoes, 85, 229. *See also* scarp retreat; spring sapping
Esplanade sandstone, 36–37
Eubrontes, *107*, 107–10, 112–17
evaporite(s), 98–99, *99*, 101, 252
evolution: of American Southwest, *143*, 143–44; of complex organisms, 4; and continent location, 115–16; of dinosaurs, 108–9, 116–18; and extinctions, 5–6, 117–18; of life, 4–6, 200–1
extension: and basin opening, 274; and normal faults, 6, 141, *142*, 237; and opening of Basin and Range, 6–7, 142, 242; and Rodinia, 3; and volcanism, 164. *See also* rift zone
extinctions, mass: of dinosaurs, 6; end of Cretaceous, 6, 117, 199; end of Permian, 5, 200, 214; end of Triassic, 4, 116–17, 215; from meteor impact, 4–6, 199–200. *See also* asteroid(s); K-T boundary; P-T boundary

fanglomerate, 18, *18*, 20–22, *21*
Farallon plate, 142, *143*, 144
fault(s), *142*; and breakup of Rodinia, 3–4; on Colorado Plateau, 35–36, 67; fast vs. slow movement on, 67; in Grand Canyon, 35–37, *36*; and groundwater flow, 265, *266*; Laguna Creek, 67; listric, 12, *13*; at Pearce Ferry, 11–17, 22–23; reactivation of dormant, 6, 67, 82, 127; San Andreas, 6, 35, 139, *143*, 144, 249; and springs, 265; Verde, *237*, 237–38, 242, 244, 249, *250*. *See also individual faults*; normal faults; reverse faults

fault slip, 11, 67
feldspar, 141, 263
Fir Canyon, 65–66
fish, endangered, 45, 48, 49–50. *See also individual species*
Flagstaff, 7, 149, 152–64. *See also* Museum of Northern Arizona; San Francisco Volcanic Field; Sunset Crater National Monument; Wupatki National Monument
floodplains: and Hermit formation, 100, 178; and Kayenta formation, 66, 113; and Mesa Verde group, 72; and Moenkopi formation, 86; and Morrison formation, 71; and Organ Rock shale, 79, *90*
flood(s): controlled, 49–50; deposits, 45, *46*, 59; and erosion, 89; flash, in Antelope Canyon, 57–58; and Glen Canyon Dam, 43; and Grand Canyon, 22; and lava dams, 34; and slot canyon formation, 57–58; and Sonsela member deposition, 206. *See also* Colorado River
folding, 224, *224*, *225*, 231, *231*
folds, 242, *243*. *See also* monocline
foliation, 222–24, *224*, *225*
footprints (dinosaur). *See* tracks
formation, 181
Fort Apache member, 181–82, *182*, *183*, 185
fossils: in Chinle formation, 203, 206, 213–16; comparison with living organisms, 96–97, 103; in Hermit formation, 100, 179; in Kaibab formation, 95–98, *96*, 102–103; marine, 99; mollusks, 103; preservation of, 118; and salt tolerance, 103; and silica, 98; in Verde formation,

fossils (*continued*)
  256. *See also* dinosaur(s);
    Moenave; petrified wood
Four Corners area, 67, *180*

gabbro, 244
gases: in pumice, 160; role in volcanic explosions, 85, 139, 153–54, 157; and vesicles, 155, *155*, 160
Giant Logs Trail, 209–12
Gilbert, Grove Karl, 191
glaciation, 163, *163*
Glen Canyon, *38*, 39–40, *43*, 63
Glen Canyon Dam, 38–51, *40*, 173. *See also* Colorado River; Grand Canyon; Powell, Lake
Gold Butte granite, 16, 21, *21*
Gowan Loop Trail, 263, 272–5
Grand Canyon, 25–37, *26*, 38–51, *41*, 93–105; age of western, 21; ancient environments in, 96–97, 99–102; vs. Antelope Canyon, 53; and Basin and Range, 22; carving of western, 6–7, 22–24; eastern vs. western, 27–29, *28*; endangered species in, 45–50, 173; erosion in, 6, 45–46, 95; food chain in, 47–48, 50, 173; fossils in, 95–98, *96*, *97*, 102–3; landslides in, 29, 95; lava features in, 25, *25*, 32–34; layers of, 11, 30, *93*, 94, 98; and river tides, 46; sandbars and beaches in, 44–47, 49–50, 173; sediment supply of, 44–45, 49–50, 173; South Kaibab Trail, 98; topography of, 95; travertine mounds in, 268–69. *See also* Grand Wash; *individual rock layers*; stream piracy
Grand Canyon Protection Act of 1992, 49
Grand Canyon Village, 32, *92*

Grand Falls, *165*, 165–74
Grand Wash, 9–12, 17, 21–24
Grand Wash Cliffs, 9, 11–12, 141
Grand Wash fault, 11–14, 16, 22. *See also* Wheeler fault
granite, 228; from Ancestral Rockies, 180; below Peach Springs tuff, 141, 145–6, *145*; at Pearce Ferry, 16, 21, *21*; near Prescott, 227, 229–30
Grapevine Mesa, 9–10, 14, *16*
groundwater: and caliche formation, 179; for Flagstaff, 163; and limestone (travertine) deposition, 254, 257, 265, 265–8, *267*; and magma, 160; from mantle, 254, 267; movement of, 69–70, *70*; and permeability, 257, 265; pumping, 76–77; royalties from, 75; and springs, 70, *70*, 133, 254, 265–67, *267*; system, 76; tables, 254–55, *255*, 257, *257*, 265

Hackberry, 141, 250
Haigler rhyolite, 263–64, *264*, *266*, 272, 274
head frames, 236, 238, 244, *245*
headward erosion, 7, 22, *23*
hematite, 181, *226*, 226–27
Hermit formation, 29, 37; absence in Little Colorado River gorge, 105; age of, 100; ancient channels in, *178*; color of, 180; depositional environment of, 100, 178–81; fossils in, 100, 179; in Grand Canyon, *93*, *94*, 95, 100–102; lateral changes in, 102, 104; and Organ Rock shale, 179; near Sedona, *178*, 178–81, *180*; source of, 179
Hickey formation, 237–38, 242
Highlands Trail, 226–29
Hitchcock, Edward, 108, 118

Holbrook, 105
Homestead Trail, 226, 229
Hoover Dam, 39
Hopi Indians: ancestors of, 122; ancestral lands of, 75–76; cultural heritage of, 65, 76; dispute with Navajo, 73–74; mining leases, 73; springs, 76; tribal economy, 65, 75–77; and water shortage, 75
hot springs. *See* spring(s)
Hualapai Indians, 50
Hualapai limestone, 14, *16*, 18–19, 22
humpback chub, 49
Humphreys Peak, 149, *156*
Hurricane fault, 36, *36*

ice age, Pleistocene, 250
impact craters, 7, 193–96, 198–99. *See also* extinctions; Meteor Crater
Indians: early mining by, 236, 252; Hopi-Navajo joint-use area, 73–74; and San Francisco Peaks, 160; in Verde Valley, 253. *See also individual tribes and cultures*
inflation, 161
iridium, 117, 199
iron formations, 3, 226–27, *226*

Jerome, 1, 225, 233–47, *233*, *236*. *See also* Cleopatra Hill; United Verde mine; UVX mine
Jerome State Historic Park, 229, 236, 238, 241, 244
Jerome volcanic arc, 1, 221, *222*, 225; collision with Prescott arc, 1, 231–32, 242–44, *243*, 274
Johnson, Warren, 45
joints, 60; cooling, 174; in DeChelly sandstone, *87*, 87–88; and dissolution of limestone, 254; and drainage patterns, 61, 63, *63*, 88; and erosion, 88–90, *88*, *90*; formation in Monument upwarp, 87; and scarp retreat, 29, *30*, 89; and slot canyon formation, 59–63, *62*
Jurassic Period, 65; and dinosaur evolution, 109, 116–18; northern Arizona during, *112*, 112–13; sedimentary layers deposited during, 71–72. *See also* Kayenta formation; Morrison formation; Navajo sandstone; Wingate sandstone

Kaibab formation: age of, 100; depositional environment of, 94, 96–97, *105*; erosion of, 98, 171, *172*; and formation of Grand Falls, 171, *172*, 173; fossils in, 95–98, *96*, *97* 102–3; in Grand Canyon, *93*, *94*, 94–98, 102; at Grand Falls, 167, 169, *170*; lateral changes in, 102–5; in Little Colorado River gorge, *103*, 103–5; at Meteor Crater, 195–97; mudstone in, 94
Kayenta, 65
Kayenta formation: at Antelope Canyon, 59; depositional environment of, 66, 69, 82; and dinosaur tracks, 107–8, 119; and Monument Valley, 83, 86; at Navajo National Monument, 66, 69–70; permeability of, 69
Kayenta Mine, 73–74, *75*, 76
Kingman, 7, 135, 146–47
knick point, 171, 173
K-T boundary, 199

Laguna Creek, 65, *68*, 69, 82
Laguna Creek fault, 67
lake(s): ancient, 7, 19, 22; Jurassic, 113; created by lava dams,

lake(s): (*continued*)
32–34; in sinkhole, 254. *See also individual lakes*; reservoir(s)
landslides: in Grand Canyon, 29, 95; in Jerome, 244; in Monument Valley, 90–91; from San Francisco Mountain, *158*, 158–9. *See also* rock fall; Sinagua formation
Laramide Orogeny: and compression, 242; and fault reactivation, 6, 67, 82, 127; and formation of the Rocky Mountains, 6, 67, 82, 126; and Monument upwarp, 82, 85; and uplift of Colorado Plateau, 6, 67, 82; and uplift of Defiance Plateau, 126–27
Las Vegas (Nevada), 3, 35
Laughlin (Nevada), 139
lava, 153; and assembly of Rodinia, 3; cascades, 25, *25*, 32; classification of, *153*; dams, 7, 32–34, 168, *169*, 170; domes, *159*, 159–62, *161*; flow, 152–55; fountains, 154; in Grand Canyon, 25, 32–34, 37; at Grand Falls, 7, 167–71, *169*, *170*, 174; and stratovolcanoes, 157; in Tonto Bridge State Park, 262, 265; in Verde Valley, 258–59. *See also* Bonito lava flow; *individual lava dams*; magma
Lava Butte lava dam, 34
Lava Flow Trail, 152–55
law of lateral continuity, 271, 276
law of original horizontality, 271, 273
Lees Ferry, 34, *38*, 42, 45
limestone formations: dissolution of, 254, 265, 267–68; in Grand Canyon, 29; by Jerome, 242; as oasis deposits, 57; and shoreline location, 97, 102–5,

265, 281, *281*; from springs, 254; in Tonto Bridge State Park, 262, 263–65, 265–68, *267*, *268*; in the Verde Valley, 7, 249, *251*, 251–54, 256. *See also* casts; Kaibab formation; Toroweap formation; travertine
listric fault, 12, *13*
lithification, 56
Little Colorado River: confluence with mainstem Colorado, 167; gorge, 102–5, *103*, 167; and Grand Falls, 165, 167, 174; lava dam on, 7, 168–70, *169*; and sediment, 44, 50, 171, 173; volume of, 165
Little Daisy mine. *See* UVX mine
Long Walk, 122

maar, 160–1
magma: chambers, 7, 135, 138, 157, 239; and faults, 37; beneath Flagstaff, 152; and gas, 85, 138–9, 154; and groundwater, 160; beneath Jerome, 239; matrix, 138; partial melting of, 163–64; silica content of, 138–39, 153, 217; from subduction, 142; and uplift, 85; variety of, 149, 163; and viscosity, 153
magnetite, 139, 226–27
Mancos shale, 72, *73*
mantle: material from, 37; partial melting of, 164; and volcanism, 7, 144; water from, 254, 267. *See also* asthenosphere
Massospondylus, 114, *115*
Mazatzal Orogeny, 3, 275, 282
Mazatzal quartzite, 3, 4, 262–69, 271–83. *See also* Christopher Mountain
Mead, Lake, 7, 9
meandering, 58, *59*, 89
Merriam Crater, 167, *168*

Merrick Butte, 86, 89
mesas, formation of, 86, 89, *90*, 251
Mesa Verde group, 72, 74
Mesozoic Era, 66; beginning of, 5; on Black Mesa, 71–72; end of, 6; sedimentary layers from, 66, 79–82
metals, heavy, 246
metamorphic rock, 3, 221–22. *See also* foliation; Mazatzal quartzite; phyllite; quartzite; stretch-pebble conglomerate
Meteor Crater, 7, 152, *189*, 189–201. *See also* Canyon Diablo; shock wave
meteorite(s), 191, *200*; at Meteor Crater, 189–201
meteor(s): effect on environment, 199–200; impact, energy of, 189; impacts on earth, frequency, 198; vaporization of, 196. *See also* extinctions; Meteor Crater
metoposaurs, 212, *213*, 214
Mid-Atlantic ridge, 5
mid-ocean ridge: and birth of San Andreas fault, 6, 142, *143*, 144; and hot springs, 238, *240*. *See also* Mid-Atlantic Ridge
mine shafts, 194, *194*, 235–36, 244
mining: at Black Mesa, 70, 73–77; environmental effects of, 8, 235, 247; and impact craters, 198; at Jerome, 233–38, 244–47; at Meteor Crater, 191, 193–96; pumice, 160; salt, 252. *See also* mine shafts; strip mining
Mississippi River, 86, 100
Mitten Butte(s), 79, *80*, 86, *87*, 89–90
Moab (Utah), 32
Moenave, 5, 106–19
Moenkopi formation, 79; depositional environment of, 79; near Grand Falls, 167; at

Meteor Crater, 193, 195–97; in Monument Valley, 86, 89
Mogollon Highlands, 6, *147*. *See also* Sevier-Mogollon Highlands
Mogollon Rim, 261
Mohave Generating Station, 76
monocline, 6; Comb Ridge monocline, *83*, 83–84, *84*; Defiance monocline, 123; and Defiance Plateau, 127; geometry of, 67–70, *68*; around Monument Valley, 82; at Navajo National Monument, 70, *71*; Organ Rock monocline, 67–70, *71*, 72–73, *73*
Montezuma Castle, 251–53, *252*. *See* Montezuma Castle National Monument
Montezuma Castle National Monument, 249–59
Montezuma Well, 249, *253*, 253–9. *See* Montezuma Castle National Monument
Monument upwarp, *83*, 82–85, 87
Monument Valley, 6, 78–91, *90*, *91*. *See also* butte(s); DeChelly sandstone; Shinarump conglomerate; spires
Morrison formation, 71–72
mountain building: during assembly of Pangaea, 4, 128; and Mazatzal quartzite, 3, 273–75, 278; and subduction, 5. *See also individual mountain ranges*; Laramide Orogeny
Muddy Creek formation, *14*, *17*, 17–23, *18*, *19*, *21*
mudstone: and erosion, 29, 84; in Grand Canyon, 29–30, 95, 100–102; as oasis deposits, 57; permeability of, 69, 257, 265; response to deformation, 87; in Verde formation, 249–51, *251*, 257. *See also* Chinle formation;

mudstone (*continued*)
Hermit formation; Kayenta formation; Mancos shale; Moenkopi formation; Morrison formation; Organ Rock shale; Toroweap formation
Munds trail, 181
Museum of Northern Arizona, 109, 159

N-aquifer, 76–77
national monuments: Canyon de Chelly, 86, 121; Montezuma Castle, 249, 251–59; Navajo, *64*, 65–66; Sunset Crater, 149–55; Tuzigoot, 246–7, *247*; Wupatki, 152, 155
national parks: Grand Canyon, 26, 94–102; Petrified Forest, 5, 124, 203–16; Zion, 114
National Park Service, 49, 123
natural bridge. *See* Tonto Bridge
Navajo Generating Station, 70, 76
Navajo Indians (Diné): ancestral lands of, 75–76; and Canyon de Chelly, 122–23; cultural heritage, 65, 76; dispute with Hopi, 73–74; and the Long Walk, 122; mining leases, 73; religious dancers, 91; song, 122; traditions, 127; tribal economy, 65, 75–77; and water shortage, 76. *See also* Antelope Canyon/Lake Powell Park; Canyon de Chelly National Monument; Monument Valley Tribal Park
Navajo Nation Museum, 109
Navajo National Monument, 6, *64*, 65–66
Navajo sandstone: at Antelope Canyon, 5, 54–57, 59; depositional environment of, 54–56; 65–66, *81*, 82; and dunes, 5, 54–56, *55*, 65–66; erosion of, 86; at Glen Canyon Dam, 5, 42–43; as groundwater source, 69–70, 76–77; and joints, 63; near Kayenta, 83; at Navajo National Monument, 5, *66*, 65–69, *68*; in Organ Rock monocline, *68*, 67–69, 71–73; origins of, 54–56; permeability of, 76; porosity of, 42; slot canyons in, 54
Navajo volcanic field, 85
Ngorongoro Crater, 139, *140–41*
Niagara Falls, 34, 173; vs. Grand Falls, 165, 167, 171, 173–74
Nininger, Harvey, 196
nodules. *See* caliche; chert
normal faults, *35, 142*; in Basin and Range, 6, 35, 144; on Colorado Plateau, 35–36; and extension, 141, *143*, 144, 237; at Grand Canyon, 35–37, *36*; in Peach Springs tuff, *140*; at Pearce Ferry, 12, 16; and volcanism, 37, 144, 164
North America: assembly of, 3, 219–21, 232; collision with Yavapai arc, 3, 274; early (proto), 1, *222*, 239, 242, 274; extension of, 3–4
North American plate, 144

oases, 57, 58–60
ocean: ancient Pacific, 4; Atlantic, 5; complex organisms in, 4; and Kaibab formation, 96–98, 102–5; Pacific, subduction under, 5; and Toroweap formation, 104. *See also* sea
oceanic crust, 1, 127, 217–19, *219*
O'Leary Peak, *161*, 162, *162*
open-pit mine, 235
ore: bodies, 235; copper, 1, 233, 235; enrichment of, 239, 242; at Meteor Crater, 193; smelting, 244–45; and sulfides, 240; uranium, 72

Organ Rock monocline, 67–70, *71*, 72–73, *73*
Organ Rock shale, 79, 86–89, *88*, *90*, 128, 179
organic matter (carbon): and chert precipitation, 98; and coal, 6, 72; and Grand Canyon ecology, 47, 50, 173; in Mancos Shale, 74; and travertine formation, 267
Owl Rock, 84, *85*. *See also* Chinle formation; Wingate sandstone
oxygen, and coal formation, 74

Pacific plate, 144
Page, 76
Painted Desert, 124, 171, 174
paleocanyon(s), *145*, 146, *147*
paleogeography, 3, 72, 100–105, *180*. *See also* paleogeography of northern Arizona
paleogeography of northern Arizona, 1–8, 271; during Jurassic, *112*, 112–13; 1,750 million years ago, 239; during Permian, 95; during Triassic, *207*
paleontologists, 114, 117–18. *See also* Hitchcock, Edward; Welles, Sam
Paleozoic Era, 4, 66; limestones, 262, 264–65, 275, *276*, *277*, 281; sedimentary layers, 4, 66. *See also individual periods*
Palisades of the Desert, 102
Pangaea: assembly of, 4; breakup of, 5; coastline, 175, 185; and Defiance uplift, 128, 131; and dinosaur tracks, 115–16, *116*; and location of northern Arizona, 4; and Navajo sandstone, 57; during Permian, *175*; and Sedona area, 175, 185
Paria River, *38*, 44, 50
parks. *See* national parks; state parks; tribal parks

partial melting, 163–64
Payson, 3, 8, 271
Peabody Coal Company, 73, 75, 77
Peach Springs caldera: erosion of, 135, 139, 146–48. *See also* Peach Springs tuff
Peach Springs tuff, 134–48. *See also* Peach Springs caldera
peaking power, 46
Pearce Ferry, 9–24
Pedregosa Sea, *184*, 185–87
perched water table, 257, *257*
periods, 1. *See also individual periods*
permeability, 69–70, 257, 265
Permian Period: in Grand Canyon, 100–105; and Pangaea, *175*; in Sedona area, 178–88, *180*, *184*;. *See also* Coconino sandstone; DeChelly sandstone; Hermit formation; Kaibab formation; Organ Rock shale; Toroweap formation
permineralization, 211, *211*
Petrified Forest National Park, 5, 124, 202–16. *See also* Rainbow Forest Museum
petrified wood, 124, 204, *208*; color of, 212; other fossils in, 212; preservation of, 209–11, *210*, *211*; root system, 212; in Sonsela flood deposits, *205*, 205–6
Phanerozoic Eon, 1, 4
Phelps-Dodge, 247
photosynthesis, 47, 173
phreatic eruption, 160, 193
phyllite, 221–23, 229–30. *See also* differential pressure
pillow basalt, *228*, 228–29, 239
Pine Creek, 8, 261–69, 272–79
plants, flowering, 212–13, *214*
plastic flow, 87

plate tectonics, 4, 217, *219*; and continent assembly, 232; and destruction of Colorado Plateau, 35; and erosion of impact craters, 198; plate reorganization, 140; and volcanism, *143*, 144. *See also individual plates*; mid-ocean ridge; subduction; supercontinents
plateaus. *See individual plateaus*
playa deposits, 20, 249
Ponderosa lava dam, 34
Postosuchus, 214, *215*
Powell, John Wesley 25, 32
Powell, Lake, 32, *33*; call to drain, 49; contaminated sediment in, 50; dimensions of, 40; and Glen Canyon Dam, 39–40, 43, 45; and Grand Canyon ecosystem, 47
power generation, 6, 39, 46, 76
Precambrian Eon, 1, 67, 128, 238. *See also* Apache group; Haigler rhyolite; Mazatzal quartzite
Prescott, 217–32. *See also* Prescott volcanic arc
Prescott volcanic arc, 1, *222*, 229; collision with Jerome arc, 1, 231–32, 242–44, *243*, 274; evidence for, 221, 226; volcanoes of, 229
Prospect, Lake, 32, *33*
Prospect lava dam, 32–34
provinces, physiographic, 9, 24, 144. *See also* Basin and Range; Colorado Plateau
P-T boundary, 200
pumice: 137, *137*, 146, 160

quartz, 42; in granite, 141; in Hermit formation, 181; metamorphosis of, 271; sandstones, 42, 79, 87; shocked, 117, 196. *See also* coesite; quartzite

quartzite, 225, 262, 271. *See also* Mazatzal quartzite
Quaternary Period, 250

Rainbow Forest Museum, 206, 209, 214
Rainier, Mount, 156, 158
razorback sucker, 49
reservoir(s), 32–34, 44, 48. *See also* lake(s); Powell, Lake
reverse faults, *142*, 230–31, *230*
rhyolite, *153*, 155, 160–61, 164, 239; bed, 272, 274–75; clasts, 272. *See also* Haigler rhyolite
rift zone, 3
ring fracture, 239–41, 244
ripple marks, 183, 280, *280*
river(s): braided, 113; channel at Canyon de Chelly (ancient), 4–5, 131, *132*; channel at Grand Falls, 169–70, *170*; channels, 86; channels by Sedona, 178–79, *178*; gravels, 262, 265, 278; Hualapai drainage, 22–23; and meanders, 58, *59*, 89, 178; Mississippi, 86, 100; Moenkopi Wash, 76; Paria, 44, 50; as systems, 46; tides, 46. *See also* Colorado River; erosion; floodplains; *individual rivers*; Little Colorado River
rock fall, 89
Rocky Mountains, 6, 67, 126. *See also* Ancestral Rocky Mountains
Rodinia, 3

Saint Helens, Mount, 135, 153, 157
salinity, 97, 104
salt deposit. *See* evaporite(s)
San Andreas fault, 6, 35, 139, *143*, 144. *See also* Basin and Range
sand, 44–45, 47, 50, 57
sandbars, 44–45, 47, 50, 173

sand dunes: and Coconino sandstone, 100–101; and DeChelly sandstone, 4, 79, 129–31, *130*; desert, 54–56, *55*, 65–66; formation of, 57; and interdune areas, 130, *130*, 133; and Navajo sandstone, 5, 54–57; and Pangaea, 57; and Sedona, 4, *184*, 185–88; and Toroweap formation, 104; and wind, 56, 183, 187; and Wingate sandstone, 82

sandstone formations: Coconino, 100–102; Dakota, 72; desert dune, 54–56, 65; and erosion, 29; in Grand Canyon, 29, 36–37, 95; in Mesa Verde group, 72; in Monument Valley, 79; at Navajo National Monument, 65; quartz, 42, 79, 87; response to deformation, 87; and shoreline location, 97, 102–5; and slot canyons, 54–57; Tapeats, 242. *See also* Coconino sandstone; DeChelly sandstone; Navajo sandstone

San Francisco Mountain, 7, 155–59, 162, *163*, 163.

San Francisco Peaks, 156, *156*, 157, 163, *163*

San Francisco volcanic field, *149*, 149–64; and astronaut training, 152; cinder cones in, 155, 167; diversity of magmas in, 163–64; lava domes in, 159–62; stratovolcano in, 156. *See also individual volcanoes*

sanidine, 137–9, *137*, 146

Sanmiguelia lewisii, 213, *214*

scarp retreat, 29, *30*; and formation of Grand Falls, 171, *172*; in Grand Canyon, 29–31, *31*, 94–95; in Monument Valley, *88*, 89, *91*; in Petrified Forest National Park, 207, 209; in sedimentary rocks, 29

Schnebly Hill formation: absence in Grand Canyon, 187; color of, 180; members of, *177*, 181–83, *183*, 185–87. *See also* Bell Rock member; Fort Apache member; Sycamore Pass member

sea (ancient): Cretaceous, 72, 74; encroachment of, 72, 100–102, 104–5, 185, 187; -floor, 1; and Mancos formation, 72; and Mesa Verde group, 74; and sediment deposition, 3–4, 66. *See also* ocean; Pedregosa Sea; sea level

sea level: and depositional environments, 66, 185; highstand, 72; and northern Arizona, 3; and Schnebly Hill formation, 185–88; and shoreline, 100, 104–5

seashells (fossil), 95, *96*

sediment: and Colorado River, 39, 44–45, 49–50; and evolution of Grand Falls, 171; and horizontal deposition, 13; in Little Colorado River, 171, 173–74. *See also* debris rafting

sedimentary rocks: on Black Mesa, 71–72; erosion of, 29–30, 95; in Grand Canyon, 27–30, 93–105; and monocline formation, 67; in Monument Valley, 79, *81*; at Navajo National Monument, 65–66; over northern Arizona, 4; passive uplift of, 67, 127; tilting of, 141

Sedona, 175–88

Sevier Highlands, 6, 71, 72. *See also* Sevier-Mogollon Highlands

Sevier-Mogollon Highlands: collapse of, 6, *143*, 144, 147–48, *147*; and Dakota sandstone, 72;

Sevier-Mogollon Highlands (*continued*)
    and location of Peach Springs caldera, 147; and Morrison formation, 71; uplift of, 5–6, 142, *143*. *See also* Mogollon Highlands; Sevier Highlands
Shinarump conglomerate, 79; at Canyon de Chelly, 5, 124–26, 128–29, 131, 133; crossbeds in, 129; and DeChelly sandstone, *125*, 126, 131, *132*; depositional environment of, 79, 86, 129, 131; in Monument Valley, 86, 88–89, *90*, 91. *See also* Chinle formation
Shiprock, 85
shock wave, 196–97
Shoemaker, Eugene, 196
Shonto Plateau, 65–77
Shylock fault, 225, 231–32
silica, 153; and chert, 95, 98; content in magma and lava, 138–39, 217; opaline, 97–98; and partial melting, 163–64; in petrified wood, 211; and volcanic rock classification, *153*, 153, 155
siltstone, 282
Sinagua culture, 154–55, 251–56, 256. *See also* Montezuma Castle; Sunset Crater National Monument
Sinagua formation, 159
sinkhole(s), 7, *253*, 253–57, *255*, 267
slag, 245, *246*
slot canyons, 5, 53–63
smokers, undersea, *240*, 240–41, *241*, 241
solar system, 1
Sonsela member, *204*, 204–7, *205*, 209
Southeast Asia, 1, *222*, 232, 239, 274

South Kaibab Trail, 98
spherules, 196
spicules, 97–98
Spider Rock, *127*, 127–8
spillway, 42–44
spires, 90; formation of, 89–91, *91*; in and near Monument Valley, 79, 84–86, 89–91. *See also* Agathla; Owl Rock; Totem Pole; Yei Bi Chei
sponges, 96, *97*, 97–98
spring(s): and evolution of life, 241; and faults, 265; and formation of Tonto Bridge, 261, 265–67; Hopi, 76; in Inner Basin, 163; in Lake Verde, 254; and limestone (travertine) deposition, 8, 254, 265–68, *268*; in Montezuma Well, 256; submarine hot, 3, 238, 240–41, *240*, *241*. *See also* smokers; spring sapping; swallet
spring sapping, 70, *70*, 133
state parks: Tonto Bridge, 261–69, 279
Steno, Nicolas 271
stratovolcano, 156–57. *See also* Rainier, Mount; Saint Helens, Mount; San Francisco Mountain
stream piracy, 7, 23–24
stretch-pebble conglomerate, 223–24, *224*
Stretch Pebble Loop Trail, 221–26; 229
strip mining, 73–74, *75*. *See also* Black Mesa Mine; Kayenta Mine
subduction, 217, *220*; and arc collision, 1, 221, 274; and differential pressure, 221; of Farallon plate, 142, *143*, 144; and formation of San Andreas fault, 6, *143*, 144; and northern Arizona, 1; plate boundary,

142; and stratovolcanoes, 156; and uplift of Sevier-Mogollon Highlands, 5, 142, *143*; and volcanoes, 1, 217, *220*, 221, 229. *See also* andesite
Sugarloaf Mountain, *156*, 156, 159, *159*, 160–61. *See also* breccia; pumice; tuff ring
sulfur: dioxide, 200, 245; emissions, 77; in lava, 154; and massive sulfides, 240
sulfuric acid, 245–46
Sunset Crater National Monument, 149–55
Supai group, 11, 102
supercontinents, 3, 115–16, *116*. *See also* Pangaea
swallet, 255
swamps, 72, 74
Sycamore Pass member, 182, *183*, *185*, *186*, 186–87

tailings, 245–47, *247*
tamarisk, 46–47
Tapeats sandstone, 242
tidal flats, 86, 99–101, 104, 182–83, 187
Tonto Bridge, 8, 260–69, *261*, 279
Tonto Bridge State Park, 260–69, 271–79
Toroweap fault, 36–37, *36*
Toroweap formation: age of, 100; crossbeds in, 104; depositional environment of, 99–100; fossils in, 99; in Grand Canyon, *93*, *94*, *95*, *99*, 98–102; lateral changes in, 102, 104; limestone in, 95, 104; in Little Colorado River gorge, *103*, 104; mudstone in, 95
Toroweap lava dam, 34
Toroweap Overlook, 25–37
Totem Pole, 79, 91, *80*, *91*
tracks: in Coconino sandstone, 188; in DeChelly sandstone, 130; erosion of, 111; as evidence of supercontinent, 115–16, *116*; interpreting, 108; at Moenave, 5, 106–19; naming of, 108; in New England, 108; preservation of, 110–13; study of, 108; swim, 114; Triassic-aged, 107; tridactyl, 107–8; underprints, 111–12. *See also Eubrontes*; Hitchcock, Edward
travertine, 261–69
tree-ring dating, 65, 154
Triassic Period: breakup of Pangaea, 5; climate, 206; and dinosaur evolution, 5, 215; dinosaur tracks from, *116*, 118; ecosystems, 117, *213*, 212–16; end of, 5; evolution during, 6; extinction events, 5, 214–15; fossils, 203, 206, 213–15; northern Arizona geography during, *207*; sedimentary layers deposited in, 79; trees of, 206, 209, *210*, 212. *See also* Chinle formation; Moenkopi formation; Petrified Forest National Park; Shinarump conglomerate
tribal parks (Navajo Nation): Antelope Canyon/Lake Powell, 52–63; Monument Valley, 78–91
tropics, 57, 179, 206
Tsegi Canyon, 65
Tuff, *137*, 137–38, *159*, 160, 225–26. *See also* Cleopatra formation; Peach Springs tuff
Tuzigoot National Monument, 246–47, *247*
Tyrannosaurus Rex, 119

Uinkaret Mountains, 37
unconformity, angular, 13, *14*, *15*
United Verde mine, 235, *236*, 244, *245*

United Verde ore body, 235; exposure of, 242; folding of, *243*; formation of, 238–41, *240*; mineralization of, 244; and Verde fault, 238
uplift, 29–30, *31*, 85. *See also* Laramide Orogeny
uranium, 72
UVX (United Verde Extension) mine, 233–38
UVX ore body, 238–41, *240*

Valentine, 144, 146
vent (crater), 154
venturi effect, 58
Verde fault, *237*, 237–38, 242, 244
Verde formation, 7, 249–51, *250*, *251*, 257; fossils in, 256
Verde, Lake, 7, 249–54, 257, 259; basalt in, 258–59, *258*; evaporite in, 252; fossils from, 256; limestone deposition in, 249, *251*, 251–52, *252*, 259; mudstone deposition in, 249, 251, *251*, 257; springs from, 254
Verde River: and Beaver Creek, 251; East, 268; and erosion of Verde Lake sediments, 250, *250*; establishment of, 249–51; and slag, *246*; and tailings, 245–47
Verde Valley: formation of, 7, 237, *250*; human history of, 235, 247; as transition between provinces, 249; and Verde River, 249–51; volcanism, 164. *See also* Montezuma Castle National Monument; Tuzigoot National Monument; Verde fault; Verde, Lake
vesicles, 155, *155*, 160
viscosity, 153
volcanic arcs, 217, 239, 274; collision of, 1, 3, 217, 219–21; in North America, 232; volcanoes of, 229. *See also individual arcs*
volcanic fields, 7, 85, 149, *149*. *See also* Navajo volcanic field; San Francisco volcanic field
volcanic necks, 84–85
volcanic rock: from ash falls and ash flows, 136; in classification and color, 153, *153*; in dikes, 227, 229–31; inclusions in, 136–8, *137*; matrix, 136, *137*. *See also* andesite; basalt; Bonito lava flow; breccia; caldera; Cleopatra formation; dacite; gabbro; Haigler rhyolite; Hickey basalt; Jerome arc; Peach Springs tuff; Prescott arc; rhyolite; tuff; viscosity; volcanic arcs; volcanism
volcanism: around Flagstaff, 7, *149*, 149, 152–64, 167–68; in Grand Canyon, 25, 27, 32–34, 37; and normal faults, 37, 144; and phreatic eruptions, 160; and tectonic plate reorganization, 6–7, 139; uniqueness of northern Arizona's, 164. *See also* ash fall; ash flow; cinder cones; lava, cascades; lava, dams; magma; Navajo volcanic field; Peach Springs tuff; San Francisco volcanic field
volcanoes: diversity of, 149, *156*; erosion of, 85, 229; explosiveness of, 85, 138–39, 153; island chains of, 1, 239, 264, 274; along subduction zones, 127, 217, *220*; volcanic necks, 84–85. *See also* cinder cones; *individual volcanoes*; lava, domes; magma; Navajo volcanic field; Peach Springs caldera; San Francisco volcanic field; stratovolcano

Vredefort crater, 198
Vulcans Throne, *27*, 32, 34

water: and coal mining, 76–77; its propensity to wander, 58; role in carving slot canyons, 57–58; superheated, 239; and the venturi effect, 58. *See also* erosion; groundwater; lake(s); reservoir(s); river(s); sea
waterfalls: and cascades, 171, *172*; and lava dams, 7, 34; migration of, *172*, 173; and travertine formation, 267. *See also* Grand Falls; knick point
Waterfall Trail, 265, 267–68
weathering, 271. *See also* erosion
Welles, Sam, 108–9

Wheeler fault, 16–17, 23
Wheeler Ridge, 11–13, *12*, *13*, 16–17, *17*, 22–23
White House Ruin, 122, *123*, 133
Williams, Jesse, 108
wind(s): from meteor impact, 189; and role in sorting sediment, 57; and sand dunes, 56, 130–31; trade, 57; transport by, 183, 187
Wingate sandstone, 82, 83–84, *84*, 86
Wupatki National Monument, 152, 155

Yavapai Apache, 253
Yavapai arc, 274
Yei Bi Chei, *80*, 91

About the Authors

Lon Abbott and Terri Cook teach at Prescott College in Prescott, Arizona. Lon is a geology professor, and Terri teaches environmental geology and water policy courses. Together they are also the authors of *Hiking the Grand Canyon's Geology*.

Lon received a bachelor's degree in geology and geophysics from the University of Utah and a Ph.D. in earth science from the University of California, Santa Cruz, where he specialized in the study of mountain building. Lon's fieldwork has taken him from remote mountain peaks in Papua New Guinea to a 15,000-foot-deep ocean trench near Costa Rica.

Terri earned a master's degree in geology at the University of California, Santa Cruz, where she studied rocks from deep-sea hot springs. Terri's undergraduate degree in archaeology is from Tufts University, and her combined interests in geology, archaeology, and experiencing new cultures have led her across six continents.

We encourage you to patronize your local bookstore. Most stores will order any title they do not stock. You may also order directly from Mountain Press, using the order form provided below or by calling our toll-free, 24-hour number and using your VISA, MasterCard, Discover, or American Express.

Some geology titles of interest:

_____Geology Underfoot in Northern Arizona	18.00
_____Geology Underfoot in Southern California	14.00
_____Geology Underfoot in Death Valley and Owens Valley	18.00
_____Geology Underfoot in Illinois	18.00
_____Geology Underfoot in Southern Utah	18.00
_____Geology Underfoot in Central Nevada	16.00
_____Roadside Geology of Alaska	18.00
_____Roadside Geology of Arizona	18.00
_____Roadside Geology of Southern British Columbia	20.00
_____Roadside Geology of Northern and Central California	20.00
_____Roadside Geology of Colorado, 2nd Edition	20.00
_____Roadside Geology of Hawai'i	20.00
_____Roadside Geology of Idaho	20.00
_____Roadside Geology of Indiana	18.00
_____Roadside Geology of Maine	18.00
_____Roadside Geology of Massachusetts	20.00
_____Roadside Geology of Montana	20.00
_____Roadside Geology of Nebraska	18.00
_____Roadside Geology of New Mexico	18.00
_____Roadside Geology of New York	20.00
_____Roadside Geology of Ohio	24.00
_____Roadside Geology of Oregon	16.00
_____Roadside Geology of Pennsylvania	20.00
_____Roadside Geology of South Dakota	20.00
_____Roadside Geology of Texas	20.00
_____Roadside Geology of Utah	20.00
_____Roadside Geology of Vermont and New Hampshire	14.00
_____Roadside Geology of Virginia	16.00
_____Roadside Geology of Washington	18.00
_____Roadside Geology of Wisconsin	20.00
_____Roadside Geology of Wyoming	18.00
_____Roadside Geology of The Yellowstone Country	12.00
_____Finding Fault in California: *An Earthquake Tourist's Guide*	18.00
_____Geology of the Lake Superior Region	22.00
_____Geology of the Lewis and Clark Trail in North Dakota	18.00
_____Geysers: *What They Are and How They Work*	12.00
_____Living Mountains: *How and Why Volcanoes Erupt*	18.00

Shipping and handling: 1 to 4 books, add $3.50; 5 or more books, add $5.00

Send the books marked above. I enclose $_____

Name_____

Address_____

City/State/Zip_____

☐ Payment enclosed (check or money order in U.S. funds)

Bill my: ☐ VISA   ☐ MasterCard   ☐ Discover   ☐ American Express

Card No._____ Expiration Date:_____

Signature _____

MOUNTAIN PRESS PUBLISHING COMPANY
P.O. Box 2399 • Missoula, MT 59806 • Order Toll-Free 1-800-234-5308
E-mail: info@mtnpress.com • Web: www.mountain-press.com